On War

New
Dialogues in
Philosophy

*A SERIES IN DIALOGUE FORM, EXPLICATING
FOUNDATIONAL PROBLEMS IN THE PHILOSOPHY
OF EXISTENCE, KNOWLEDGE, AND VALUE*

Series Editor
*Professor Dale Jacquette, Senior
Professorial Chair in Theoretical Philosophy
University of Bern, Switzerland*

In the tradition of Plato, Berkeley, Hume, and other great philosophical dramatists, Rowman & Littlefield presents an exciting new series of philosophical dialogues. This innovative series has been conceived to encourage a deeper understanding of philosophy through the literary device of lively argument in scripted dialogues, a pedagogic method that is proven effective in helping students to understand challenging concepts while demonstrating the merits and shortcomings of philosophical positions displaying a wide variety of structure and content. Each volume is compact and affordable, written by a respected scholar whose expertise informs each dialogue, and presents a range of positions through its characters' voices that will resonate with students' interests while encouraging them to engage in philosophical dialogue themselves.

On War

A Dialogue

Brian Orend

ROWMAN & LITTLEFIELD PUBLISHERS, INC.
Lanham • Boulder • New York • Toronto • Plymouth, UK

ROWMAN & LITTLEFIELD PUBLISHERS, INC.

Published in the United States of America
by Rowman & Littlefield Publishers, Inc.
A wholly owned subsidiary of The Rowman & Littlefield Publishing Group, Inc.
4501 Forbes Boulevard, Suite 200, Lanham, Maryland 20706
www.rowmanlittlefield.com

Estover Road
Plymouth PL6 7PY
United Kingdom

British Library Cataloguing in Publication Information Available

Library of Congress Cataloging-in-Publication Data:

Orend, Brian, 1971–
  On war : a dialogue / Brian Orend.
    p. cm. — (New dialogues in philosophy)
  ISBN-13: 978-0-7425-6044-4 (cloth : alk. paper)
  ISBN-10: 0-7425-6044-9 (cloth : alk. paper)
  ISBN-13: 978-0-7425-6387-2 (electronic)
  ISBN-10: 0-7425-6387-1 (electronic)
  1. War—Moral and ethical aspects. 2. War—Philosophy. 3. Peace. 4. United States—
Military policy—Moral and ethical aspects. I. Title.
  U22.O63 2009
  172'.42—dc22
                              2008028152

Printed in the United States of America

⊚™ The paper used in this publication meets the minimum requirements of American
National Standard for Information Sciences—Permanence of Paper for Printed Library
Materials, ANSI/NISO Z39.48-1992.

For Barry Turk, *a.k.a.* "Bumpa"

Table of Contents

Acknowledgments

Thanks first to Dale Jacquette for inviting me to take part in this exciting series, wherein I was required to stretch my writing skills and to add to my experience. Thanks, too, to the helpful and patient people at Rowman & Littlefield, especially Ross Miller, Doug English and Alan McClare.

Thanks to the University of Waterloo, and all my students both past and present, for making my job such a joy. Special thanks to Debbie Dietrich, who gave me good advice and managed somehow to decipher my increasingly flawed handwriting and type the entire manuscript in record time.

In 2007, I got to tour the major U.S. military academies and speak about postwar justice. Thanks very much, for making this possible, to Martin Cook at the Air Force Academy, George Lucas at the Naval Academy and both Dan Zupan and Todd Burkhardt at West Point.

As always, deepest thanks are reserved for my friends and family, especially Krista, Mom and Barry (to whom this book is dedicated). Huge thanks to my beautiful boy, Sammy, for all the fun we've had bonding over football this past year: Go Giants Go! Finally, thanks to Jennifer McWhirter for all the love, and just for being around, gracing and gilding my life.

Dramatis Personae

Dramatis Personae (in order of appearance):

Lori Gordon: Gil Gordon's wife, a master's student at Columbia University in NYC

Gil Gordon: Seriously injured U.S. soldier, a Major in the U.S. Army

Nick Gordon: Gil's uncle, a one-star U.S. Army General

Elizabeth McAllister: U.S. Army attorney and Captain, investigating Gil's injuries

Ty Leung: Gil's doctor and surgeon

Carter Johnson: U.S. Army Corporal, master's graduate from Oklahoma State

David Pearson: Professor of philosophy, U.S. Military Academy at West Point

Diego Cortez: U.S. Army chaplain, ex-Catholic priest

Chapter One

Injury and Empire

Lori Gordon got out of the taxi, showed her I.D. to the guard, and was escorted—quickly, grimly—to the U.S. Army hospital room where her husband lay unconscious, severely injured on the battlefield in Iraq.

She was exhausted, from the long flight from America to Germany and then the land trip from the Frankfurt airport to the U.S. Army base where the hospital was. This base has been here since the Second World War, she said to herself, as she looked about her husband's room. Looks like it hasn't been cleaned since, she thought. She spoke aloud: "Uncle Sam strikes again."

Her bright blue eyes turned to her husband, Gil, who was almost unrecognizable: heavily bandaged and hooked up to all these machines, tubes, monitors, and IVs. He seemed as though he was also strapped down to the huge, oversize hospital bed.

Lori exhaled, and pulled both hands through her fine, shoulder-length hair. It used to be red, but recently had been dyed blonde. She tucked her hair behind her ears, and felt her face flush. Her husband's thick brown hair was only half visible; the other half of his head, and most of his face, was wrapped in big bandages that had some blood and ooze seeping through them.

It wasn't good, Lori knew. When she got that call on her cell—that dreaded call, from a faceless Army official—she nearly screamed and smashed the phone onto the sidewalk. But, she kept her composure until she had come home, booked the first flight out, and nibbled at some much-needed food. She had always feared precisely this moment—why hadn't he listened to her?—and now here was Gil, fighting for his life while strapped down to a bed. And in Germany—they only flew the most serious cases out of Iraq and up to the big base in Germany. And then they called to tell her she'd better come . . .

Suddenly, into the room strode a one-star U.S. Army general, and a very tall, very attractive woman also in uniform. They were both holding cups of coffee, and seemed to be in quiet, awkward conversation.

The general—alert, rigid and focused, with salt-and-pepper buzzed hair and chiseled features—exclaimed: "Lori!"

"Hi, Nick," Lori replied while walking up to the general, giving him a hug. He kissed her quickly on the cheek. He was Nick Gordon, her uncle-in-law, eldest brother of Gil's father.

The general looked at Lori straight in the eyes, gripped her with both hands firmly, but gently, on her upper arms, and then stood to the side slightly while gesturing towards the tall woman.

"Lori, this is Elizabeth McAllister. She's an Army attorney. There's some question as to how Gil got hurt—whether it was friendly-fire or enemy-fire— and Command wants a thorough investigation right from the start. Just met her myself, we were having coffee."

Elizabeth offered her hand to Lori, who shook it. The former said: "I'm sorry to have to meet you like this, and for the injuries to your husband."

"Thank you." Lori noted in Elizabeth the sharp feminine features, the freckles, the bright blue eyes, and the voluminous head of curly, dark red hair. Lori resented the height difference between them. Then she asked: "There's a chance this was caused by Americans, or Allies, accidentally hitting Gil? Not Iraqis or terrorists?"

Elizabeth replied frankly: "Apparently it's possible. They as yet have no firm idea, the 'fog of war' and all. But in my experience the fog clears after a good investigation and so I've been assigned to play a part. If Gil's injuries were friendly, there might be grounds for disciplinary action or court martial, and so they need legal people right from the start. The general here has been assigned to escort me from D.C. down to the theater in Iraq to look at evidence and to interview all the players, but we wanted to stop here first to see Gil and you."

"Ties of family," the general interjected.

Lori: "Thank you, Nick. It is very nice to see family. So does anyone know what happened?"

Nick: "No, I'm sorry. We weren't fully briefed on the first flight but will be—so they say—on the second flight down to Iraq. I'm the one who dragged Captain McAllister down here straight away, from the Pentagon, once I heard the gravity of Gil's injuries."

Lori asked, trembling hand wiping away at the start of some tears: "And so what is wrong with him?"

Nick: "Considerable head trauma. Head and back of neck. Like shrapnel from a roadside bomb, or bullets. They looked at him on-site and sent him up here. Stuff's lodged in his brain, maybe some in his spine. Too delicate and difficult to do down there. So, here we are."

Lori: "Have they done anything on him here?"

Nick: "They did exploratory surgery, plus all the usual—MRI, and so forth—to get the best view of what they're going to have to do. The doctor said a big concern is brain swelling in response to the presence of foreign objects. They've essentially put him into a coma to handle that."

Lori: "When will they do the surgery to get the stuff out?"

Nick: "It all depends on the results of these findings. The doctors must bring it all together, then debate among themselves the next steps."

Lori: "No timeline?"

Nick: "They couldn't give me one."

Lori looked at Elizabeth, who shot her a look of sympathy. Then Lori asked the lawyer: "Soldiers can be court-martialed for mistakenly hitting one of their own or one of our Allies?"

"If it was truly an accident, no," Elizabeth replied. "And the chaotic nature of battle means that accidents happen. But there is a legal principle that, if the 'accident' was so bad—sloppy—that it was essentially negligence on someone's part, then they must be held accountable. Negligence means the person responsible failed to act as any other reasonable person, or normal soldier, would have under the same circumstances. That's what I'm going down there to determine. The U.S. military likes to thoroughly investigate all such allegations and keep everything ship-shape."

Lori let out a snort of ridicule. "Ship-shape, huh? I'd think that'd be a Navy expression. Ship-shape with regards to these investigations, but happy to be involved in a horrible, utterly unjust, and completely stupid war to begin with. Why aren't you investigating Bush for the crime of starting this war?"

Elizabeth was shocked at the outburst.

"Lori," the general cautioned.

Lori: "No, Nick, no. Don't tell me to back off, or to calm down. There's Gil! There's Gil. Right there! Twenty-five years old, married to me for four years. Twenty-five, Nick! Now with half his head blown off, or a spray of bullets in his head, or whatever. And for what, Nick? What? And why? What if he dies? What if they can't do the surgery, and my husband dies and 'poof,' there goes all our dreams for a happy life?"

Nick: "Gil believes in the security of America, Lori."

Lori: "Yes, he does, Nick. Maybe he shouldn't, but he does. He came to believe in patriotic values like that partly because of you, as you know. You urged him to enlist."

Nick: "I did, and I still would. The Army gave Gil—and you, Lori—some opportunities that otherwise wouldn't have happened. College. A secure job . . ."

Lori: "Secure job! Look at him! Secure job . . . .Yeah, you are guaranteed your position until you get killed—possibly by your own side—and then we'll re-assign your 'guaranteed position' to the next poor schmuck who comes around."

Nick: "Lori, I won't hear you bad-mouthing the military like this. Especially not in an Army hospital."

Lori: "Why not? Because you've made it your career, and it pays your wage, and you've been lucky enough—and it's just pure luck, Nick—not to get shot up like Gil?"

Nick: "It's not just selfish, Lori. I believe in the defense of America. America is my home. I believe in defending my home. I also believe that America is the greatest country on Earth, and is a force for good in the world, and that it's a good life being able to contribute to that."

Lori turned on Elizabeth: "Do you believe that patriotic garbage, too? You wear the uniform—did you swallow the same pill of patriotism as my Uncle Nick here?"

Elizabeth flushed, and then looked down. She should have realized family members might be hostile. But she'd never been exposed to this, and didn't think it through. Why hadn't she flown straight on to Baghdad?

Lori demanded: "Well?"

Nick interjected: "Lori, she's not family. You can't talk to her that way. We've had similar discussions before. Captain McAllister's just doing her job, and is only here because I dragged her. I've got to chaperone her—Army orders—but wanted to see Gil first."

Lori: "And this might be the last time you see him alive, Nick. And let her an-swer for herself."

Elizabeth replied in a soft, measured tone: "I don't want to cause problems, or to argue with you. But I'm in the Army for a reason. It has given me great legal training, which I hope to use in the future."

Lori: "Inside or outside the Army?"

Elizabeth: "Outside."

Lori: "Higher pay?"

Elizabeth: "Yes, but that's only one reason."

Lori: "The others?"

Elizabeth: "Well, let's not get too personal, please, but the military is still a very male-oriented world and I feel there are limits to what a woman can realistically achieve, whereas I can use my training and experience here and take it into private practice, where I hope I can earn a very decent living and then settle down and have kids."

Lori: "You sound a lot like Gil did, when I begged him not to join up. Just thinking of the benefits, never the costs. And kids . . . kids . . . ," she paused, then flared: "And you know, I want kids, too! And there's the guy I want to be their father! Look at him!"

Elizabeth: "I'm very sorry."

Lori: "Well, no, look, I'm the one who's sorry. I'm tired and in shock and angry. My husband's all shot up and has shrapnel in his brain. And I've just always disagreed with people like Nick, and now my worst fears as a wife are coming true . . ."

Elizabeth: "It's OK, it's very understandable."

Lori: "But, if I do understand you, you're not the same as Nick. You're not in this for patriotic reasons, but just personal ones."

Elizabeth: "Yes, but I am grateful for the opportunities which the military has given me. My investigation skills are light-years ahead of my nonmilitary peers."

Lori: "But you're still going to get out, because it's a macho world filled with patriotic jocks."

Elizabeth: "I wouldn't say it quite like that, but I do plan on leaving. It's not as woman-friendly as it might be."

Lori: "Or minority-friendly."

Elizabeth: "Well, I wouldn't know about that . . ."

Lori: "Why, because you're white? Because you're white, you can't see how the military exploits nonwhites, and especially African-Americans?"

Elizabeth: "Colin Powell rose to the top."

Lori: "Well, he's just one man. Let me tell you something. When I think of the U.S. military, I think of it as a protection force—like a private security force—for the elites of the U.S. Empire. It provides the muscle—the poor, minority muscle—to protect the wealth, power, and interests of people who are already the richest and most powerful on Earth. Old, rich, white, Protestant men—the very men who've always ruled this country, and for their own benefit."

Elizabeth was appalled, and didn't know what to say.

Nick interjected: "Paranoid nonsense, Lori. I'm sorry, I'm trying to hold my tongue here out of respect, but can't. You're being rude. The U.S. military is made of free citizens—men and women, white and black, Asian and Hispanic—who freely choose to help defend the greatest nation in history."

Lori laughed: "OK, so let's really get into it. You don't believe that America is an empire—like Rome in its day—and that the U.S. military is the world's police force, which protects and expands this empire, and that the main beneficiaries of this activity are the American aristocracy?"

Nick: "Aristocracy?! That went out with the French Revolution, Lori, and has never been part of the American experience. And neither has empire. That's what Europeans, and other cultures, do—or have done. You simply cannot say that America is an empire. Indeed, it was founded in violent revolution *against* a European empire: England. America was founded on great moral values—individual rights to life, liberty, and the pursuit of happiness—as opposed to the older, much more cynical vision of a European empire, which was just about power; which group got what; which group was on top; no higher moral purpose whatsoever."

Lori: "You're wrong. Wrong on both counts. You'd have to be blind not to see that there is, in fact, an American aristocracy. Old, rich, white, Protestant, powerful, men. You don't think there are social classes in America? And that some have way more power—economic, legal, and political—than others?"

Nick: "But there's no official entrenched aristocracy, like in old European empires. Look, of course there are inequalities—that's the price of living in a free society. America remains the land of opportunity. If it weren't, it wouldn't keep attracting so many immigrants."

Lori: "Well, it's interesting you mention immigrants, Nick, because that's relevant in a different way."

Nick: "How so?"

Lori paused to explain something to Elizabeth: "I'm writing my master's thesis on the idea of an American empire."

Elizabeth: "Really? Where?"

Lori: "Columbia. Gil had been stationed at West Point. Anyway, Nick, as I was starting to say, on the relevance of immigrants: every empire in the world has existed on the backs of slave labor."

Nick: "There you go, Lori, we don't have slavery. There was a little thing called the Civil War, which ended that. Remember a guy named Abraham Lincoln?"

Lori: "Well, Nick, there's slavery, and then there's slavery, right? And we did have slavery to begin with, as no one can deny. And today, we have tons of immigrants—legal and illegal—who do the slave-like jobs, for starvation wages with no benefits, which are essential to the U.S. economy. Yard work. House cleaning. Nannies. Dish-washers. Fruit-pickers. A clear laboring underclass, just like all the old empires. And the fact that we attract so many just fits right into our status as the mother country, the core at the hub of the empire."

Nick: "What? You're losing me, Lori."

Lori: "C'mon, Nick, its simple. Every empire in the world—Greek, Roman, Spanish, French, English—has featured a fundamental split between the mother country (or the core or the hub, as it is also known) on the one hand, and on the other, the colonies (or dependencies or hinterland or periphery). The core rules the colonies for its own benefit, sucking resources out of them and making itself richer. I mean, have you seen Paris or Rome? The core sends its own people to govern the colonies, imposes its own values on them, but it's all about growing the core at the expense of the periphery. America has all the classic signs of a core: the world's biggest economy, attracting most of the foreign investment; it is also a magnet for immigrants desperately seeking a better life. America has the most influential culture—TV, Hollywood, all the celebrities—and is, politically and militarily, the most potent country by far. You know the French have a new word for us?"

Nick: "What do I care what the French think?"

Lori: "Well, the word is 'hyperpower.' We're not just a superpower anymore, Nick. That was back during the Cold War with the Soviets. No, today . . ."

"After we won that war," Nick interrupted.

Lori: ". . . yes, today, we are the one and only 'hyperpower.' The global hegemon. Niall Ferguson has written a book on us called *Colossus*."

Nick: "Well, why shouldn't he? We *are* the global colossus. Aren't the most dominant nations always a hegemony? All those things you said about America's influence? They are true. For sure. But they don't add up to the conclusion that America is an empire the way Rome or England was."

Lori: "Sure they do."

Nick: "No, they don't. America is the most powerful country on earth—true—but that doesn't make it an empire. People just accuse it of being that because they are jealous and resentful of American power. Being powerful and being an empire are *not* the same thing."

Elizabeth jumped in: "I'd have to agree, Lori. Legally, when you look at previous empires, there was a clearly defined—formally, legally defined—status of dependency and inferiority, which all these colonies had vis-à-vis their mother countries. All the empires knew, and declared, that they had empires, and they legally defined the subservience of their colonies. They specified what they would take from their colonies, and what they offered in return, which was usually a swap of economic resources and manpower in exchange for military security, governance, and investment in their development."

Lori: "But there's *formal* empires, and there's *informal* empires. True, America has not declared itself an empire, nor does it admit it has one. But that doesn't change reality. As I've described it, it's crystal clear that America has a real empire. As you yourself just said: America sucks up a ton of resources, is the richest and biggest economy, has the most influential culture and foreign policy and, in exchange for having all this stuff, invests in foreign countries and gets involved in policing the entire world with its military because its interests are now global. America is an empire, just an empire-in-denial. In fact, it's been an empire even from the first moment of its very existence."

Nick: "What? Baloney, Lori! Typical conspiracy theory. Maybe we have very recently had some foreign entanglements, like in the Middle East, but it wasn't always that way."

Lori: "Don't be so sure, Nick. Again, as with Elizabeth, it all depends on how you define and perceive 'empire.' Think, for a second, about how this country was first founded: the Pilgrims landed on Plymouth Rock in 1620, seeking freedom from the religious persecution they suffered back in England. And who did they find in the earliest colonies? Indians, Native Americans. And the European settlers proceeded to conquer them, driving them off their lands. If you look at a map charting the historical growth of the United States, you see this relentless spread and expansion: northward and southward, but especially westward—always westward—until the present day United States was created. Right from day one, the elites in America engaged in a consistent policy of imperial expansion; taking land from the Indians, driving them away or killing them, bringing over a ton of slaves from Africa to do the work, and then they grew fat and rich and powerful themselves. And it only stopped when they ran out of land, at the shores of the Pacific Ocean."

Elizabeth: "If you admit that it was stopped."

Lori: "No, I don't. Sorry, I misspoke. It actually didn't end at the shores of the Pacific. From there, America hopped over to Hawaii, and jumped up to Alaska. I suppose it has always been interested in Central and South America, the Monroe Doctrine and all that. Then it went over to West Germany and Japan after World War II. You know, we ruled them directly—imperial control doesn't get more direct than martial law, that is, direct military control with no local government—for years and years after that war. And now we're in the Middle East. And why?

Because the American Empire runs on oil. We need the oil to manufacture things, heat our homes and offices, and of course, to drive our precious, precious cars and trucks and vans and SUVs. I was talking to a friend the other day who said that we've reached the height of Roman corruption and decadence; we go to war in the Middle East to get the oil we need to fuel our NASCAR races, which we run simply to entertain ourselves."

Nick: "Outrageous, Lori! First, we are not in the Middle East to get the oil."

Lori: "Well, we're not there to get all the sand!"

Elizabeth laughed, but Nick continued: "We are in the Middle East to defend and secure the American homeland from terrorism and horrible, outlaw regimes which fund and support terrorists, start wars, and try to develop dangerous weapons that we can't let them have."

Lori: "So we can have them and they can't?"

Nick: "That's right. Because we're good, and they're bad. If they were stable, decent, good governments, it would be a different story. We don't demand the British or the French give up their nukes, but we can't responsibly let terrible governments—like Saddam Hussein's in Iraq, or the Taliban in Afghanistan—gain that kind of weaponry. It's not hypocrisy, Lori; it's just good common sense, and the ability to distinguish between good countries and bad. Speaking of which, so what if America is an empire? Let's say I was to agree with you, just for the sake of argument; at least America is a *good* empire."

"Is it, though?" Lori asked, smirking.

Nick: "Sure it is. Look, America is not perfect and, yes, parts of our history are flawed and sad. Human beings make mistakes. But America still stands for great values: individual rights; property; democracy; separation of church and state. America beat the Nazis, then the Soviets. America faces down threats from fascists and communists and others who don't believe in the value of the individual, or the worth of human freedom."

Elizabeth: "And I would support that, Lori, and also point out that, if you look at the world, and see what has happened to it since America truly became a world power, it really looks as though America has benefitted the world: everyone is so much richer; there is so much more knowledge, culture, and technology; there's less bigotry; and America really has defeated these major threats to decent living. I mean, the global economy has just boomed since America won the Cold War in 1990, and capitalism triumphed over communism. So, isn't America a good empire?"

Lori: "Well, can there really be such a thing? I mean, an empire is all about power and control. It's about exploiting people and not letting them have a say, or local control, over their own lives. If we really believed in freedom and democracy, wouldn't we have to say that all empires are necessarily unjust?"

Nick and Elizabeth paused. Lori continued: "But you know Ferguson takes up this point in his book. He says America *could* be a good empire, but currently *isn't*. America could be like the British Empire, and provide protection and security for the world (or at least its colonies). It could invest in these colonies; help them become more advanced and developed. It could share these great values, like human rights and democracy. It could, but it doesn't. Instead, America denies it even has an empire. As such, it refuses to take on its proper responsibilities to those over whom it has control. You know, back in the day, the best and brightest of the British aristocracy would compete over who got to govern a given colony. When they got there, they would have it as a point of pride and ambition to develop their colonies to the best of their ability. You think that is true today of America's best and brightest, America's next generation of leaders? Are they striving to become, say, President of Iraq, bring in peace and growth to that country? No, they never want to leave America: they either want to go to New York (to become rich on Wall Street), go to Washington (to win political power), or go to Los Angeles (to become famous in Hollywood). Hell, let's face it, they know very little about the world outside America. They neither know nor care. They are therefore lousy imperial overlords. No wonder Iraq is screwed up."

Both Nick and Elizabeth drained their coffees at the same time. Nick then spoke while looking into his empty cup: "That might be true, if we *were* truly an empire, which we're *not*. And that might be why we know comparatively little about other countries and cultures: we don't have to, because we are not, as you claim, 'their imperial overlord.' These people must want, and learn, to govern themselves. America is not the world's keeper, or the global policeman."

Lori: "And yet here lies my husband—originally from Colorado—at this U.S. base in Germany, for head injuries sustained in an American war in, and occupation of, Iraq."

At that moment, a doctor arrived.

Chapter Two

Doctors and Definitions

The doctor was older: maybe late '50s or early '60s. He was Asian-American, of shortish height, and bursting with muscles. He obviously hit the gym very regularly. He wore his black hair very trim, and had a reflective air about his wide face which seemed slightly at odds with the rest of his body.

Nick immediately turned and shook his hand. "Doctor," he greeted, with a crisp nod.

"Hello, General," the doctor replied with an eye on the general's insignia. "I'm Ty Leung. Pleased to meet you."

"Likewise. I'm Nick Gordon, Gil's uncle. This is his wife, Lori, and this is Captain McAllister."

Dr. Leung: "Nice to meet you all. Lori, I'm sorry to tell you that Gil is in quite bad shape right now. We've induced a coma to prevent further brain swelling."

Lori: "Nick told me."

Dr. Leung: "Good. There are many pieces of shrapnel and fragments lodged in his face, neck, and brain. The surgical team is piecing together the evidence, and we are mapping a strategy."

Lori didn't reply.

Dr. Leung continued: "His vitals are stable for now, but the injuries sustained are severe. We are going to take our time mapping the surgical strategy, since we not only have to remove the foreign objects, but to do so in the least invasive way, and then to be very careful about sewing everything up after we've removed them. Gil may well be looking at a series of surgeries over several days."

Lori looked out the window.

Dr. Leung: "Our staff are arranging lodgings for you here on the base. A junior officer will be by to brief you, and look after you, in this regard. I just wanted to meet you, on behalf of the surgical team, and let you know where we're at."

Nick: "How long can he stay in a coma like that?"

Dr. Leung: "Provided no new bleeding starts or that the shrapnel doesn't substantially shift, we can sustain that for many days. But we don't want to, we don't think. We're hoping at least to have the strategy mapped out by the end of the day, or tomorrow at the latest, and then to proceed with the first surgery. We don't want those things inside his head for long. They can shift, and shred tissue or nerves, provoking bleeding."

The doctor looked at Captain McAllister: "And, I'm sorry, Captain, I didn't catch why you are here."

Elizabeth: "I'm an attorney from the Pentagon. We don't know how Gil and his unit were injured. Apparently there's a chance it was friendly-fire. I'm here with the general, en route to Iraq to investigate."

Dr. Leung: "I see."

Elizabeth: "Is there anything you can tell me about Gil's injuries that might shed some light?"

Dr. Leung: "Well, I'm not sure yet. Do we need to write a report about that?"

Elizabeth: "Yes."

Dr. Leung: "OK, we'll keep all the foreign objects for analysis as to their origin. There's quite a few of them, which might suggest shrapnel from a roadside bomb, perhaps even one loaded with nails."

Lori finally spoke: "What?!"

Dr. Leung nodded his head: "That would be consistent with some of his injuries. The enemy packs an explosive device with nails. The nails survive the blast and get sprayed out at incredible speed, maximizing casualties."

Nick: "If the blast doesn't get you, the flying nails will."

Dr. Leung: "Exactly. Very cheap, but very deadly. The evidence is pointing in that direction, though I suppose a number of pieces might resemble bullets."

Lori: "How many pieces are there?"

Dr. Leung: "Over a dozen."

Nick: "How many big ones?"

Dr. Leung: "About half that. The smaller ones are actually more problematic. You have to dig further, and they're harder to get out." The doctor paused. "They don't know how the injuries came about? No one saw a roadside bomb go off?"

Elizabeth: "I'm sure they did, it's just that we haven't been briefed about it. And so, I gather, neither has your medical team."

Dr. Leung: "No, and it would be important information. Could you please try to get that to us?"

Nick: "I'll get on the horn straight away."

Dr. Leung: "Thank you. I would also appreciate knowing if more like Gil are on the way up. Was he the only casualty of the incident?"

Nick: "Again, we've had no briefing."

Dr. Leung: "OK, whatever you can do would be appreciated."

The doctor turned to Lori: "We've had maybe ten cases like this in the past three months, Mrs. Gordon. We know what we're doing. This is how the enemy fights, and it produces casualties of this nature. We'll do everything we can for Gil."

"Thank you," *Lori replied. Then she added:* "How did those ten cases turn out?"

Dr. Leung took in a breath: "Well, keep in mind only the most complex injuries get flown up here, and so I'm sorry to say that, if memory serves, only about four survived."

Lori looked at the doctor with a blend of anger and sadness.

Dr. Leung: "These injuries are very hard to treat. The number of shrapnel, the shredding they do. The enemy fights in a nasty way."

Nick: "Because they can't, or won't, come out in the open and stand tall, fighting like men."

Lori: "Well, it wouldn't be smart for them to do that, right, Nick? If they did that, U.S. forces would blast them apart in no time."

Nick: "And put them out of their misery."

Elizabeth: "They want to live to fight another day. So they don't engage directly. They do sneaky, indirect targeting—and with messy weapons, like bombs filled with nails—and achieve their goals without direct confrontation."

Lori: "And what are their goals? To drive the United States out of Iraq, at minimal cost to themselves. It's cowardly, but it's rational. And Gil's now a victim of their strategy."

Dr. Leung: "Mrs. Gordon, I wish I were still of fighting age. I'd be down there myself, fighting with my band of brothers, sticking it to the enemy."

Nick: "Amen to that."

Dr. Leung: "On the one hand, I'm a surgeon now and I help that way. On the other, seeing and treating all these young American boys makes me furious. The rest of the surgical team agrees. The enemy fights like a coward, and the price is paid in American blood."

Lori: "Another patriot."

Dr. Leung: "Sure, I'm a patriot. I love America. I believe in defending the homeland from attack. I'm a surgeon, a scholar, and a warrior. I model myself after Socrates."

Lori: "After Socrates, the ancient Greek philosopher?"

Dr. Leung: "Yes. My hero, a true hero. Died for his principles, defending the laws of his land."

Lori: "I didn't know he was a surgeon, or a warrior."

Dr. Leung: "No on the surgery—that's my own thing—but yes, he served in the Athenian army, like all able-bodied citizens did. We should have mandatory military service today in America, like they do in countries like Israel. Helps bring a country together."

Lori: "But why should a country be 'brought together'? If you ask me, it is precisely all this patriotic, love-of-country feeling that causes wars in the first place. And the military instills it in you all because that serves its own objectives and keeps you in line, gets your fighting spirits up, makes you do dumb, 'brave' things on the battlefield which serve the unit."

Nick: "You really believe that patriotism is the root cause of war?"

Lori: "Sure, it's pure tribal loyalty. Herd behavior, exploited by the leaders of the herd for their own benefit. If only I could have persuaded Gil of that. I mean, the doctor just said only forty percent of similar cases he's treated recently have survived. That's a nice way of saying there's a sixty percent chance Gil will die. And for what? What good does it do *him*? Or *me*? Or the *children* we hoped to have? All to serve the tribe, out of a completely insane and irrational identification with the pack."

Dr. Leung: "A scholar, Michael Gelven, in his book *War and Existence,* says that the essence of war is precisely about 'us versus them.'"

Lori: "And he's right."

Dr. Leung: "But he seems to suggest that's good."

Lori: "Then he's wrong. He's correct about war being a fight between human tribes. Clearly, we're tribal, either by nature or by nurture. But how can he possibly think this is good? War is good? Like, what, does he own shares of stock in all these weapon manufacturing companies or something?"

Dr. Leung: "He says that war brings about a group solidarity, and common purpose, that no other activity can match. He says war is the ultimate affirmation of a people's life and its values. War is the surest way we assert the values of our way of life—of who we are—over that which is not ours, that which is strange and alien."

Lori: "Well, it's his thoughts that seem strange and alien to me. Pure war mongering, in my view. A ridiculously romantic perspective on war. Totally male, too—glorifying the killing of the other, and the assertion of the self."

Nick: "Wait a minute, wait a minute. We're running a whole bunch of topics together here. But this *is* interesting. Let's put some thought into it: there's the issue of *what war is*—how to define it completely and accurately—and then there's the separate issue of *what causes war*, whether for instance it's patriotism, or too much testosterone. Or all those weapons manufacturing companies."

Elizabeth: "And then there's a third issue still—and a big one—*whether war is good*, or whether it can be legally or morally permissible."

Nick: "Let's start with the issue of defining war. You know, I'm a general. I saw combat action in Vietnam, Central America, and Iraq. Yet I've always found this basic, conceptual issue tricky. But what I can do is quote Clausewitz."

Lori: "Who's Clausewitz?"

Nick: "Former officer—general, I think—back in the old German state of Prussia. Lived during Napoleon's time, if I'm not mistaken. Wrote a book called *On War*. Everyone used to have to read it. Mainly about strategy and the need for speed. Anyway, I'll always remember that he said that 'war is the continuation of policy by other means.'"

Elizabeth: "That is supposed to be a definition? It's so vague."

Nick: "It is rather philosophical, but those Germans—our hosts here—have been known to be that way. But think about it: he's saying that the point of war is to achieve the aims of policy when other means of achieving them aren't enough."

Elizabeth: "That only leaves me asking for more. I thought good definitions were supposed to answer questions, not provoke further ones."

Nick: "Well, hey, let me finish. Let me continue my train of thought. What does policy try to do?"

Lori: "What do you mean by 'policy'? Whose policy?"

Nick: "The government's. Why does government exist? To rule. To rule a people in a given land. Policy, I think, refers to the way the government wants to rule. For instance, the government makes a policy to grow the economy. Then, it needs to decide on the means to achieve that policy, like cutting taxes or negotiating a trade deal with a foreign country."

Lori: "OK. So what?"

Nick: "When peaceful measures of policy fail, war is always an option. At least on international issues. Say you're having a problem with a foreign country. First, you try diplomacy to talk it out, craft a deal, negotiate a settlement. If that fails, and the issue is very important, war's always the final option for getting what you want."

Elizabeth: "But how does that tell us *what* war is?"

Nick: "It tells us that war is a tool for governance. War is the use of armed force in order to help a government rule its people, or to help solve a dispute over ruling with a foreign government. War is a violent way of ruling."

Elizabeth: "Well, there's no denying that. But then there's the issue of defining violence."

Nick: "The use of force."

Elizabeth: "And force?"

Nick: "You know, physical force. Like using physical strength to force someone else to do something they don't want to do. Using your body to inflict physical damage on someone else."

Elizabeth: "Just physical force? Physical damage? There's no such thing as psychological violence? What about verbal abuse?"

Nick: "Well, verbal abuse between individuals isn't relevant to war between countries, is it? We can't get sidetracked. How does the law define violence or war?"

Elizabeth: "The law says violence is the unlawful use of physical force. So, it's not 'violence' unless the force violates some community standard establishing a difference between good force and bad force."

Nick: "So physical force can be either good or bad, in the eyes of the law?"

Elizabeth: "Of course."

Nick: "Just like with war: some wars are good, like the Second World War; others are bad, like the First World War. But the essence of war is the fact that it is the use of physical force to achieve a policy aim of a government. Clausewitz had this great further definition, something like 'war is an act of violence, like a duel, designed to compel your opponent to fulfill your will.' See, it fits in perfectly with what I've been saying: war is the use of physical violence between countries to settle a policy dispute they are having."

Elizabeth: "International law adds detail by specifying that the 'physical force' on display is, more concretely, the mobilization and deployment of armed forces—that is, the army, navy, air force, marines, coast guard, what have you.

Ordering them into military action across a border is the essence and start of war."

Dr. Leung: "I can add to this. Gelven says that war is *violent, vast,* and *political.* So, we've got a good general idea now of violence—this notion of using physical force to try to inflict your will on someone else, harming them in the process—but you forgot to add that the quantity of force must be vast, or huge, or big. It's war, after all, not a fight between gangs on a street corner on a Saturday night."

Nick: "Good point. But how big is big? How vast does the force have to be for it to count as war?"

Dr. Leung: "I'm not sure. Does there have to be a precise calculus? All I'm saying is that war is not a fight between individuals, like a boxing match, nor between small groups like criminals, gangs, or cops and robbers. It's a big, *widespread* outbreak of violence."

Nick: "OK."

Dr. Leung: "And then the political aspect comes into it. It's not just violence, breaking out at random. It's done *for a political reason.* I think here's a good connection to the whole policy thing you mention from Clausewitz—the reason for the fighting, and sending of troops and weapons over a border, is to secure a policy objective. And policy is about ruling or, um, about organizing a people and its life in the land on which it lives. The intention is political. It's not about economics, or personal hatred, or group rivalries, or a fight between men over a woman, or a 'turf war' between criminal gangs over controlling the drugs in a neighborhood. *It's about trying to shape the social life of a whole people in a given territory.*"

Elizabeth: "But it's not just that the intentions, or motives and aims, are political in that way. It's that the very actors in motion in war are themselves political bodies, or entities. One thing I don't like about our discussion so far is that it doesn't account for civil wars *within* countries. We've been talking only about national armed forces crossing international borders with a political objective in mind; classic interstate wars, like the World Wars or the Persian Gulf War in 1991. What about the American Civil War, or the civil war that tore apart Yugoslavia in the 1990s?"

Lori: "Or the civil war happening right now in Iraq?"

Elizabeth: "Well, let's leave Iraq to the side for now. But the point remains; surely these conflicts count as wars even though, technically, the violence is all within one country and it never crosses a border."

Lori: "What might help here is Max Weber's old distinction between nation and state."

Nick: "Weber?"

Lori: "Old, dead, German sociologist. Again with the Germans. Oh well, here we are in *Deutschland*, right? Anyway, Weber said that a state is strictly just the governing structure of a society. The state equals the government. And the state, as you were sort of saying Nick, tries to govern its people in a given territory. Weber actually drew very close links between governing and violence, saying that one of the things which defines a state is that it is the only agency in society with a monopoly—one hundred percent control—over the legitimate use of violence. You know, the police force, army, navy, and so on. Of course, others—like criminals or an abusive husband—might still use violence, but only the government has *the right* to use violence. And it has the right because it is needed to establish law and order in a given society. Society would just fly apart—collapse into anarchy—without one 'neutral' agency in charge of making rules and policies, and then having the right to use violence if necessary to enforce them."

Elizabeth: "What does this have to do with civil war?"

Lori: "Getting there, sorry. But before I finish with Weber, I did want to mention this clever book by Bruce Porter, where he draws a direct historical link between the rise of the modern state and warfare. He wants to draw a dark lesson: not just the obvious one that wars influence governments but that, in his view, governments have used war, and the cloak of war, to expand their control over their own societies. Classic example: income taxes were introduced as a 'temporary measure' during the First World War. They are still with us today, feeding the resources and growth of state control over our lives."

Nick: "We're getting offtrack, Lori."

Lori: "OK, sorry again. Though, in fairness, that's completely in keeping with your line about war being the continuation of policy by other means. Anyway, back to Weber. The state is the government. It governs a people. Now, a nation is a group of people which views itself as a clear, strong, and separate group with a distinct way of life. It does this based on things like language, religion, culture, fashion and dress, cuisine and cooking, um—what else?—a roughly common worldview and values, shared historical experiences, and even ethnicity and skin color and things like that. A nation is a group which thinks of itself as a separate people."

Elizabeth: "What about the term, 'nation-state'?"

Lori: "Well, that's just it. Many states govern over many different nations. But modern history has shown the potency of the idea of the nation. Most nations want their own state. They want to govern themselves, since they see themselves as distinct. Hey, talk about causes of war, nationalism has got to be a big one. How many wars in history have been fought over nations striving to become, or to have, their own state? And here, finally, is the connection to civil war: sometimes, a state will fall apart, and the nations inside will fight over control of the central state, or they will seek to separate from the central state and establish their own, newly independent state. Like America did to Britain in the War of Independence, or the South tried to do against the North during the American Civil War. Conclusion: we don't say war is a thing which only happens between countries or states. I propose we say it happens between nations, too."

Dr. Leung: "OK, but what about the war on terror? The war which is the reason why Gil is injured, and which started on 9/11. You can't really say that terrorist groups are nations, can you? Yet, still, we are at war with them."

Lori paused. Then Elizabeth spoke: "Well, it is hard to see how al-Qaeda counts as a nation. Nor is it a state. But, obviously, it would like to become a state—love to gain control of a government—and then use the powers of that state over time to create a nation. Control over the education system, trying to enforce a religion through the legal system."

Dr. Leung: "Maybe what we're looking for, then, is a kind of term like 'community,' or 'political association.' The relevant actors in war are *political associations*—groups of people with a clear political agenda. 'Political' means power, power over people; power to organize the lives of people in a given land. Clearly, terrorist organizations are political associations, in that they have this as their agenda—they want power, they want to realize their dreams of an ideal society. Ideal to them, anyway. So, it's acceptable to say they are belligerents during war."

Elizabeth: "And using 'political associations' also fits with civil wars since nations warring against each other within a single country, or over one single state, are clearly political associations. They are even wider and deeper and more robust political associations than terrorist groups, which can be just a band of individuals scattered across several countries with nothing more in common than a radical political and/or religious doctrine. Nations, in addition to common views, speak a common language, share a cuisine, and so on. I guess I'm saying we can define 'political associations' along a spectrum: with established governments or states, at the fullest extreme, and moving through nations, civil war factions, and then terrorist groups at the other end of the extreme."

Lori: "I like it; 'political associations' works because it still excludes individuals and nonpolitical associations, like criminal gangs, or business corporations, or religious institutions. War happens for *political* reasons between *political* actors."

Nick: "So, where are we in terms of our definition?"

Elizabeth: "Well, let's see: war happens between political associations for political reasons—that is, for reasons having to do with struggling over who controls the state or policy in a given land. War is the widespread, or vast, or substantial use of physical violence between political associations."

Nick: "Good. I like it."

Dr. Leung: "Another addition."

Nick: "Let's hear it."

Dr. Leung: "The use of force must be *deliberate* and *intentional*. Better yet, it must reflect the overall will of the political association in question. I say this, for instance, thinking of India and Pakistan."

Elizabeth: "What about them?"

Dr. Leung: "Well, they are suspicious of each other. And they have serious disputes about the territory of Kashmir. The British Empire used to run India. When it left, in the late 1940s, it felt that, for social peace, it had to split India into a mainly Hindu India and a mainly Muslim Pakistan. But where Britain carved out the border between those two new countries was very controversial, and the Kashmir zone remains disputed territory to this day. Wars have been fought over it. My point is this: quite often, the border patrols from each country will take shots at each other, or hotheaded, ambitious officers will throw a missile or two at the other side to see if they are awake or asleep. That's the use of force across a border, between political associations, but we don't want to say that counts as war."

Elizabeth: "That's where the *quantity* or *vastness* of armed force comes into it."

Dr. Leung: "Yeah, OK. I see that. But I'd still suggest a friendly amendment to our definition."

Elizabeth: "If there are no objections," she paused and looked quickly at the discussants: ". . . then I'd say, that leaves us with the following: *War is the intentional and widespread use of physical violence between political associations.* Physical violence in general is the use of force to inflict damage upon another and to try to compel them to do your bidding and, specifically, physical violence in connection with war is the use of armed forces—that is, guys with guns—in pursuit of a policy objective. Finally, the term 'political association' means a group of people seeking to substantially affect the policy of a state."

Nick: "Excellent. We ain't not dumb." He smiled.

Lori: "But we should stress the territory part of it."

Elizabeth: "How so?"

Lori: "Because, in the end, power is exercised *somewhere*. People have to live *somewhere,* and it's neither the sky nor the ocean. It's on land. War is a fight over people and over land. It's about power *over* people *in* a given territory. This all gets back to my comments about tribalism: War is about which tribe owns which piece of land, and controls which other tribes in the neighborhood. And, in today's world, the 'neighborhood' is all of Earth. So, war is all those things you said, but in connection with the goal of using the violence to gain or defend territory and/or to control or crush other tribes."

Nick: "Well done, everyone. But now what about the causes of war? I mean, in our discussion about definition, how many possible causes of war did we come up with? I mean, let's list them."

Lori: "Let's see . . . tribalism, group rivalry, desire for territory, nationalism, patriotism—are those last two the same?—testosterone, weapons manufacturing companies . . ."

Dr. Leung: ". . . the lust for power, a perverse love of violence, bad border-drawing by colonial powers, religious differences."

Elizabeth: ". . . but we should add economic reasons, like controlling oil and natural resources . . ."

Dr. Leung: ". . . and what about just good, old-fashioned ruthless scheming by ambitious politicians, who want to grow their own territory, get more resources, maybe consolidate their power at home or to distract their people's attention from a problem back home . . ."

Lori: ". . . and let's not forget just plain stupidity and miscalculation, and yet not backing down, which leads to a fight."

Nick: "Wow. That seems like everything and the kitchen sink. Speaking of kitchen, I'm hungry now. It's past lunch hour. Let's get some chow and continue discussion while we eat."

Dr. Leung: "You all go ahead. I've got to rejoin the surgical team. General, you did promise me information on how Gil was injured, and whether more are coming."

Nick: "That I did. Sorry, I got caught up in our conversation." He took out his cell phone. "I'll do that right now. Ladies, give me five minutes, and I'll escort you to the mess hall."

Lori looked at her husband's body. Dr. Leung read her expression. He said: "It's OK to leave him. He's stable. Severe, but stable. Have a good lunch. I'll report back later."

And with that, they all left the room.

Chapter Three

The Causes of War

Lori: "So they still don't know exactly how Gil got injured?"

Nick: "Well, a whole unit was involved. They were on patrol in Baghdad. Their vehicle *did* run over a bomb in the road, and right afterwards they came under fire. Translation: it was an ambush. Gil may have been hit *both* by shrapnel from the bomb and by bullets during the shoot-out."

Elizabeth: "So no chance of friendly-fire?"

Nick: "There is a chance. Apparently, the firefight was chaos—having just been hit by the bomb—and some guys may have been hit by their own. It happens. Fog of war. Firing in panic at what you think is the enemy."

Elizabeth: "Doesn't sound like negligence, though. If friendly-fire happened, it would seem as if it were a sincere accident. That would make my life both easier and happier."

Lori looked down into her food and flushed.

Elizabeth: "I'm very sorry, Lori. That was not the best thing to say. I'm sorry."

Lori didn't acknowledge the apology, and Elizabeth quietly cringed. She looked at Nick and he just shook his head, signaling: "Let it go."

The three were finishing lunch in the base's enormous mess hall, and were the only group around, everyone else having eaten and departed long ago. Lunch was tomato soup, salad, and chicken sandwiches, washed down with coffee and orange juice. Everyone thought it was surprisingly tasty and filling, for mess hall chow.

Lori: "We might as well get back to our conversation. I found it very stimulating. A good distraction from Gil, at least for a few minutes anyway. Let me kick things off, this round, by saying that war is caused by the military-industrial complex. Iraq is all about oil and profits. Dwight Eisenhower himself—ex-general and ex-president—warned back in the fifties about this. His predictions have come true."

Nick: "No. This war is about protecting America from terrorism, and saving Iraqis from the tyranny of Saddam Hussein."

Lori: "Well, Saddam's gone now. I'll give you that. And I can hardly deny he was a brutal dictator. But really, Nick, do you honestly think the average Iraqi citizen is clearly better off now than before the invasion and the overthrow of Saddam? Before, it was admittedly a dictatorship. But at least there was law-and-order. Or order, anyway. Now, there's so much violence and chaos, and their economy is in shambles, and there's all these different groups and factions fighting for power."

Elizabeth: "But they've had free elections, and a new constitution."

Lori: "True, true. Those are real achievements, especially considering Iraq's history. But are those things enough to satisfy the average Iraqi? How do such abstract things, like democracy and more freedom, translate into concrete benefits, like peace and more money in their pockets? That's the key, in my view, to postwar success over there. They've got to feel better, postwar, than they felt prewar. And I don't think they feel that way, and so what was the war for?"

Nick: "You're too impatient, Lori. Like way too many other people back home in America. Postwar reconstruction takes a lot of time. We can't simply give up and go home. They need us over there. The whole thing would collapse if we weren't there."

Lori: "Correction, *Nick:* the whole thing collapsed because we went in there, and kicked it over. And for what? To get control of the oil. So the massive oil companies can get even more massive, and all these American private security companies over there—protecting the oil companies—can make a ton of money, and all the weapons manufacturing companies can make a killing (ha ha) peddling their wares to everyone: to the U.S. military; to U.S. private security companies; to the new Iraqi army and police force; and then even to the insurgents themselves. The weapon-makers win no matter who wins. It's all about who profits, those are the people who push for it. Greed. Greed is the ultimate cause of war."

Nick: "Wow, Lori. That's all just so paranoid. I think I'd go crazy if I believed the same things you do. In terms of the weapons manufacturers, they produce a product and sell it to those who freely demand it. They don't cause or create wars."

Lori: "But they profit from them."

Nick: "True but, again, it's not like these companies themselves start wars. That's the responsibility of political leaders."

Lori: "That's too superficial, Nick. You have to dig deeper and ask yourself: 'Why do political leaders go to war?' It's obvious that they go to war to get their hands on things for their friends and countrymen back home. The American car driver, and home owner, needs oil. American oil companies want more money. American security and weapons companies want more profit, too. And when the companies make more, they hire more workers, giving more Americans more jobs. So you order a war to make all these things happen. It's 'us versus them.' I still remember TV coverage from when the first Persian Gulf War started, in 1991. U.S. troops were leaving one of the bases in the South. The news reporters showed images of families along the roadside cheering the soldiers on as they drove by. One of the cheering fans, so to speak, held up a simple handwritten sign. It read: 'Kick their ass, and get their gas!' That guy—a simple guy in T-shirt and jeans, with a beer belly and a baseball cap—got it. He understood the essence of war. Resorting to violence to get stuff for ourselves. No better, no different—in fact, much worse, when you think about it—than children on a playground."

Elizabeth thought to herself: "Again with the children." Then, she spoke: "But Lori, maybe—with respect—that thesis is too simple and superficial. Your bringing up the 1991 Iraq War made me mindful of a book that was published around that time: Samuel Huntingdon's *Clash of Civilizations*."

Lori: "I've heard of it, but never read it."

Elizabeth: "Very interesting, and very controversial at the time. In it, Huntingdon speculates about the essence of war, and what causes it."

Nick: "And what does he say?"

Elizabeth: "He says the very opposite of Lori's thesis. Lori's thesis is inspired by Karl Marx, the founder of communist philosophy. For them, it's all about the economy. It's all about material goods and natural resources, economic production, and competition. War is conflict over these things, over who gets to own and control them, and to profit from their use. Lenin—the leader of the first communist revolution in Russia—went even farther, to say that the logic of capitalism creates war."

Nick: "How so?"

Elizabeth: "Because there are only so many natural resources—land, water, people, oil, gold, timber, and so forth—to go around. They aren't making any more land, in particular. So Lenin thought this would drive capitalist countries to even more intense levels of competition between themselves, over these things. Eventually, this would lead to war. He suggested that greed produced slavery and colonialism, and would only spark bigger and more ferocious wars over time. He wrote at the time of the First World War."

Nick: "So war is a form of economic competition?"

Elizabeth: "Yes, for Marx and Lenin."

Lori interjected: "And me."

Elizabeth smiled: "And Lori, too."

Nick: "So there's no solution to the problem of war, as there will always be greed and economic competition?"

Elizabeth: "Some may say that. But Lenin, for one, thought that, if we got rid of capitalism, we'd have our solution. He didn't believe that humans are greedy by nature. We're greedy because the capitalist structure of our society makes it advantageous for us to be greedy. The key for him, and certainly for Marx, was allowing private property ownership of the economic means of production, like natural resources. When people are allowed to own and accumulate things, that's what they'll do. As a result, you'll have a greedy, self-absorbed, 'me-first' culture filled with conflict."

Nick: "So the solution is not to allow anyone to own anything? This will solve war?" He snorted in disbelief.

Elizabeth: "Something like that. If we created a different kind of society—a communist one, with no private property allowed, and with equal sharing of economic resources—then we would behave differently. Think about Lori's analogy to children fighting over things on the playground. If no one were to own anything—if everything was public property—then they wouldn't fight over these things. There'd be nothing to gain."

Nick: "But communism failed. So they must have been wrong."

Lori: "Just because communist regimes collapsed doesn't mean communism as an idea is wrong. Many so-called 'communist' societies were just military dictatorships under a different name. Besides, the capitalist countries were keen on defeating communism, for the sake of their own greed. There's no guarantee that having the best idea will make you prevail in the real world, the world of dog-eat-dog."

Elizabeth: "Ah, your reference to 'the best idea,' Lori, gets me back on track to Huntingdon. Back in '91, the Cold War was ending. For decades, we had this tense, indirect struggle between capitalist and communist countries. Between America and the Soviet Union in particular. America wins because, from 1989 to 1991, all the communist regimes in Eastern Europe, Central Asia, and the Soviet Union collapse, and get replaced with free market democracies. Capitalism wins."

Nick: "Yay."

Lori: "Boo."

Elizabeth: "Huntingdon, at that moment, asks himself: 'what is the future of armed conflict? Where was the next battle to be fought? Who would be the next big enemy?'"

Nick: "And what did he say?"

Elizabeth: "He said the next big enemy, for the West, would be found in the Islamic world."

Lori: "What?! He said that, back in '91? Before 9/11?"

Elizabeth: "Ten years before."

Nick: "Fascinating. He predicted the future correctly."

Elizabeth: "Well, at the time—as I mentioned—his thesis was greeted with massive hostility. He was even labeled a racist by some, a warmonger by others. But his ideas have stood the test of time. Which brings me, finally, to my point. For Huntingdon, war is a clash of civilizations (hence the title of his book). It is a conflict of ideas. A battle over values: moral, political, philosophical, and even religious values. Maybe even especially over religious values and spiritual doctrines. I guess, in general, that war is a violent disagreement over whose values and beliefs should prevail in a given land, or amongst a given people. So, this is why I said that Huntingdon's thesis is the exact opposite of Lori's: for her, war is over concrete, material resources; for him, war is over abstract, immaterial ideas and values. Especially over ideals regarding what makes for a just society."

Nick: "What does that have to do with the West and Islam? We're reliving the Crusades, is that it?"

Elizabeth: "Not exactly. But your reference to the Crusades is one Huntingdon would use to support his thesis. Just think, he would say, of all the wars in history fought over religious ideals, and whose vision of God should prevail. The Crusades, from 1000–1300 AD, were fought by Christian Europe to stop the spread of Islam, and ideally to recapture Jerusalem from Muslim control, so as to pave the way for the second coming of Christ."

Lori: "What? That's what the Crusades were fought over?"

Elizabeth: "I think so, at least initially."

Lori: "C'mon, there must have been other reasons, like political power, or economic gain. Someone must have made some money, or thought they could."

Elizabeth: "There must have been weapons-makers back then, too, I guess. But Huntingdon would agree with Nick: these people were *responding* to demand rather than *creating* it. And the demand came from religious and political leaders, trying to impose a worldview, a comprehensive culture, or at least a vision of how a society should be run . . . along which lines, according to which ideas and whose values. Another example he likes to use is the Thirty Years' War, in Europe, from 1618 to 1648. This was a terrible war over religion—between Protestant and Catholic—in Europe. The Protestant Reformation, starting in 1517, ended the monopoly of the Catholic Church over religion in Western Europe. The

Catholics launched their 'Counter-Reformation' and, before you knew it, they were all going at each other in a series of wars. The worst was the Thirty Years' War, which had a particular affect on Germany (and here we are, right?). One-third of Germany's population dies in this war. Can you imagine that—easily, worse casualties than the American Civil War. And over what? Over religious differences. Not money or resources. Ideas."

Nick: "And that would even apply to our own civil war. I haven't heard anyone make a compelling case that economics or greed played a big role in our civil war. For me, that was about two huge ideas, or political values: preserving the Union (you know, saving democracy); and abolishing slavery. I like this Huntingdon guy."

Lori: "C'mon, *Nick:* don't drink the Kool-Aid. Of course, political leaders are always going to cover-up their greedy, nasty, and base motives with glorious sounding rhetoric. No, this war can't possibly be about money; it's about freedom, or ending slavery, or spreading democracy, or the truth of Christianity. That's how they get the gullible public to buy into the war and participate. But who benefits from the war, Nick? That's the real question. Is it John Q. Public? Is it the average soldier? No, never. It's the big decision-makers, and the weapons-makers, and all their friends and associates. In other words, as Eisenhower said, the military-industrial complex."

Nick: "But the military-industrial complex didn't exist back during the Crusades, or the Thirty Years' War, or the American Civil War."

Lori: "Not by name and of course the Industrial Revolution doesn't happen until 1750 or so. But you're missing the point, Nick. The point is that you can't take what these leaders say at face value. They can be—and perhaps often are—liars and hypocrites. You focus your attention on who benefits from these things and, if that is the standard, it's clear who's behind war. *Get rid of the business of war; that's how you get rid of war.*"

Nick: "How are you supposed to do that? Armies always need suppliers. Wars usually produce winners (as well as losers). How do you change that?"

Lori: "As Elizabeth was saying, by changing the structure of society. Look, for example, at what they did here in Europe. After WWII, some bright bulbs here in Germany, and others in France, said: We can't ever have another war between us. We can't afford it and, now that the world has nuclear weapons, the prospect of yet another war between us would be terrible. Who knows where it would end? So, they decided to link their two economies together, and so tightly that the prosperity of one would be directly linked to the well-being of the other. It would literally be economic suicide now for the one to attack the other, given the enormous growth in interdependence between them, which has been the basis for the European Union."

Nick: "But now you're saying that business is a good thing? I thought you said 'the business of war is the cause of war'? So, which is it?"

Lori: "Look, those were two different points; one on cause; the other on solution. I still think the military-industrial complex is the cause of war. But my point with France and Germany was that the solution to the problem of war lies in changing institutions, and the connections between societies."

Elizabeth: "That's interesting, Lori. But, before we go there, I don't think we've given Huntingdon's thesis a fair shake. Let's go back and finish it off before returning to your perspective."

Lori: "OK."

Elizabeth: "Huntingdon argues that the only way to explain how so many people are willing to make so many sacrifices over the years—the sacrifices needed to sustain a serious war—is to appeal to their sense of justice. Greed just won't cut it: few people are willing to risk their lives for their greed. Whereas more are willing to die for the sake of their ideals, especially their political and religious ideals."

Nick: "He's right."

Lori: "No, he's not. This is my whole point about political rhetoric."

Elizabeth: "Anyway, back in 1991, he wrote about the ideals of the West, and he looked around at the other major civilizations in the world—African, Asian, Indian, Oriental, Russian—and concluded that the Islamic world was committed to the values most opposed to those of the West. Hence, it would be the next big enemy, especially since the West and Islam rub against each other in Turkey, the Middle East, and elsewhere. There are flashpoints for conflict. And that's how he sees the oil. It's just a spark for the conflict; but what keeps the fires of war burning is the deeper clash of values. So, Huntingdon predicted a decades-long new Cold War, so to speak, between the West and the Muslim world."

Nick: "Impressive. Two questions, though. What exactly is 'The West'? And what exactly are the Western ideals he's talking about? Make it one more: What exactly are the opposed Muslim values he mentions?"

Elizabeth: "The West refers to Western Europe, and all the countries it created through colonialism. So, geographically, you start at Greece—where it all begins—and go west to the Atlantic coast. And you throw in Britain, too. All those countries within that space are Western, and, through colonization, many more are, too: all of the Americas, Australia, and New Zealand, and so on."

Lori: "Are you saying India is a Western country, because the British used to run it as a colony? Didn't they also run Kuwait, and lots of Africa? Are those places Western, too?"

Elizabeth: "No, I don't think so. Because Western societies didn't *create* those countries. Those societies already had an independent existence for centuries, and various Western powers just conquered them and ruled them for awhile.

Whereas Western countries created nations like America, Australia, and Canada."

Lori: "What about the Caribbean countries? Or Japan, clearly an Asian country, yet reconstructed in its modern form by America following WWII?"

Elizabeth: "That brings me to the point of Western values. This is really the crucial thing about the West: it's a state of mind. Specifically, Huntingdon lists the following core values of a Western society: individualism, and the commitment to respect individual human rights; free market capitalism, where private property ownership is allowed; government is seen as the servant of the people, and there is a deliberate effort to control the size and growth of the state, for instance through checks-and-balances, a written constitution and, most crucially, free, fair, and regular public elections. There is democracy and the rule of law: even government officials must adhere to the same rules as the rest of us. There are other groups and associations permitted apart from the state—political scientists call this 'civil society'—and also, very importantly, there is separation between church and state. The government does not try to realize a religious vision through its laws and the power of the state. Finally, in the West, there is a commitment to the use of science and technology to improve people's lives. There aren't religious-based obstacles standing in the way of science for the sake of preserving some traditional doctrine."

Nick: "What about Christianity?"

Elizabeth: "What about it?"

Nick: "Well, call me dumb, but how can a society possibly be called 'Western' without being Christian? Not in the church-style sense—of the government implementing Christianity, but in the sense that the socially prevalent religion is, or was, Christianity."

Elizabeth: "Well, I'm not sure that's needed. I consider myself a Western person but not a Christian. Religion doesn't play any role in my life."

Nick: "But we're talking about Western civilization, Elizabeth. The whole thing and not just you. Surely one of the things that define Western civilization is Christianity."

Elizabeth: "Maybe you're right, or at least you once were. I'm not sure anymore. Many Western countries—not so much America as over here in Europe—are now being described as 'post-religious.' And I say 'Amen' to that."

Lori: "I'd say 'Amen' to that, too. But I don't think we can deny the historical truth of Nick's point."

Elizabeth: "Maybe."

Lori: "One more thing I thought of, for the West, or what defines it: the role of cities, growing urban life, at the expense of the rural countryside, at least since the Industrial Revolution."

Elizabeth: "I think Huntingdon, or one of his followers mentioned this. Cities as the economic, social, and cultural engines. Thanks."

Lori: "No problem. Continue."

Elizabeth: "Now, keep in mind these are all ideals. This is not to say these ideals are all perfectly realized in the West, nor to deny that there are things like lying, hypocrisy, and weakness of will, wherein you fail to live up to your own ideals."

Lori: "To say the least . . ."

Elizabeth: "But, if you look at this list of Western values, and those of the Islamic world, at least as it was in 1991, you will see almost complete disagreement. The Muslim world is not Christian, obviously. Free market capitalism was not allowed; Middle Eastern economies are very heavily regulated by the government. There are few, if any, checks on government power, and civil society activity is not robust. In 1991, I don't think there was a single Muslim democracy, nor one with the impersonal rule of law (as opposed to the personal rule of dictators or religious leaders). Speaking of religion, there isn't a clear separation between church and state in many Muslim countries and, in some like as Iran, the very point of the government *is* to realize Islamic principles through the power of the state. I think they may appreciate some science and technology—especially weapons-based technology and technology which helps them extract oil—but it's clear there is religious-based hostility to technology, too. The birth control pill, I'm told, is hard to come by, for instance. And that raises the whole issue of women and individualism. Women do not have the same status as men in the Islamic world, and Muslim nations consider themselves more social—and family-oriented than what they perceive as a very selfish and isolating culture of individual happiness in the West."

Nick: "And cities?"

Elizabeth: "That, I don't know for sure. In some Islamic countries, there are huge cities, like Cairo or Tehran. But I also think, that, in terms of cultural ideals, traditional Islam views city life as unhealthy, decadent, and not communal enough."

Nick: "So that's a point-by-point refutation of Western values, and an assertion of Muslim ones."

Elizabeth: "Yup. The key ones, I think, have to do with individualism and the power of the state. We believe in every person's having—at least in principle—human rights to life, liberty, and the pursuit of happiness. We think the state can be a danger to this, given our history. So, we want to curb state power. Whereas, over there, there is more comfort with state power, perhaps because there is a more social view of life straight away. Islam means 'surrender to God,' or 'submission to God,' right? Submit yourself, as an individual, to the power, supremacy and command of God. Not exactly permission to pursue happiness however you see fit."

Lori: "No, indeed. But then what was his conclusion again about oil, in the context of the current war?"

Elizabeth: "That it is this enormous gap and disagreement over values, and how to run a society in particular, which, uh, fuels this war. Oil is just what brought our two civilizations together. Though, I guess, religion brought us into conflict long before that. Oil is just the spark, the thing igniting the fire. But what will, and is, keeping it burning is *this difference in what we each believe as civilizations.* Just like the Cold War between capitalism and communism, and destined to last as long: a decades-long fight for 'civilizational' supremacy. War is motivated by the desire to see one's ideals prevail."

Nick: "I've never believed this nonsense that Iraq, and America's presence in the Middle East, is about oil. I mean, sure, we need oil. We're an advanced economy with many machines and manufacturing. And God knows we love to drive. But look, we're also the richest country on Earth. We can buy all the oil we need at market prices. They'll freely give it to us in exchange for our money. So who the hell needs the burden of running an oil colony? Plus, if we were after oil colonies, we would've kept Kuwait after we saved them from Saddam in 1991. Instead, we restored them to independence. And we are trying to do this in Iraq, but it's just a lot more difficult because of the violence, and the fighting between the groups there. And, if we really wanted oil colonies, we would also be trying to take over Nigeria, Venezuela, and Saudi Arabia. And I actually heard Canada and Russia have a ton of oil, almost as much as the Saudis. Why aren't we trying to take them over? Why settle for dinky little Iraq when we could have Saudi Arabia, or Canada? With Canada, it's just a quick drive north to Alberta, a few days worth of work. No, we targeted Iraq and Afghanistan because they were rogue regimes which threatened the national security of the United States. We are in the Middle East to protect ourselves from another 9/11, and to help protect our allies like Israel."

Lori: "I'm not convinced. Anyway, let's move on to consider other theories on the causes of war. What about feminist theories, which focus on men as warmakers?"

Nick: "Oh, here we go . . ."

Lori: "No, Nick. You can't dismiss this with a roll of your eyes. It's a real theory which deserves consideration."

Nick: "Does it? OK, let's hear it and we'll decide."

Lori: "Well, men like to fight. Men love competing against each other—in sports, in business over money, in politics over power. Men fight each other over social prestige, over promotions, and over who gets which woman. It's all so ape-like, in the end. We've not escaped our monkey-like heritage and we know, from everywhere in the natural world, that the alpha males can't stand each other, and they do what they need to—including vicious fighting—to duke it out, and see who prevails. It's the law of the jungle: it's dog-eat-dog, with the alpha males leading the struggle over territory, money, resources, ideas, values, and even women. Maybe even especially over women, in the end. I mean, that's the case in the natural world, or the animal kingdom, right? It's all about which male gets to breed with the most desirable female, right? It's just that, with humanity, the competition between men is more complex than in the animal world, and it's not just brute strength, speed, and physical features, but also about money and power, social status, and real estate, and so on. But it turns out that even those things can be had through physical force, too: just a more complex deployment of physical force in war."

Nick: "So war is about getting laid?! War is about sex?! Get real, Lori. Few things, trust me, are un-sexier than being in battle and fighting for your life. That's like saying, I dunno, that being in a car crash turns you on, which is just plain sick."

Lori: "What the hell are you talking about? Car crashes?! That's a bad analogy. Hear me out, and open your mind. Especially to the things which can be learned about human behavior from animal behavior."

Nick: "Lori, we're not animals, OK?"

Lori: "Yes we are."

Nick: "Well, yes, we're an animal species, part of the natural world, and we may or may not be descended from apes."

Elizabeth: "What? Of course we are."

Nick: "You believe that. I believe man was created in God's own image."

Elizabeth paused: "Wow."

Nick: "And I'll not apologize for that belief. Anyway, even if we are just glorified monkeys, and God's got nothing to do with it, the degree of that glorification—the sheer extent and magnitude of greater intelligence, ability, complexity, and morality that we have over and above even the greatest animal—renders these comparisons really suspect. So, it's ridiculous and insulting to say humans fight for the same reasons monkeys do—to get sex and then to get lots of bananas for all the babies who follow the sex. We are way more gifted and complex and we fight for more complex reasons, like Huntingdon and the ideas and values thesis."

Lori: "You have a lot of confidence in our giftedness and our differences."

Nick: "Why shouldn't I? I always laugh when I hear of all the psychology experiments with all the rats, and then how these scientists make these crazy generalizations from rats to people. Boggles my mind. Science it ain't. These people just like playing with rats, and they justify their existence by pretending there's a practical application to humans."

Elizabeth: "Ok, but let's not lose sight of where we are, with this feminist thesis."

Lori: "Yes, Nick, think about it. Who have always been the major figures in war? Men. The major leaders in wartime? Men. The vast majority of soldiers? Men. The main scientists behind weapons development? Men. The main CEOs behind weapons companies? Men. The main war criminals and war heroes? Men. Men, men, men, men, men. This cannot simply be coincidence. Testosterone has been linked to aggression. Maybe there's your connection between war and sex—the same hormone which fuels the male sex drive also fuels man's propensity for physical violence."

Nick: "Is that true about testosterone? I thought adrenaline was behind spurts of physical activity and violence: the fight-or-flight response to perceived threats. And women have adrenaline, too."

Lori: "I'm pretty sure it's testosterone."

Nick: "And I'm pretty sure you're not a biologist."

Lori: "Look, chemicals aside, you still can't deny the correlations I mentioned between men and war. Men are fascinated with war. Who is buying all these war books, and watching all these war movies and TV documentaries? It's not the local ladies club. You men are fascinated by violence, and are much more likely to resort to violence when there's conflict. Its how you work out your anger, whereas women have other ways."

Nick: "Better ways, I suppose? Like crying and knitting, or shopping?"

Lori: "Shut up, Nick. And look at the obsession men have with military technology and the latest weaponry. Can it be random chance that all these high-tech missiles look like huge penises? That the ultimate weapon itself—a long-range, intercontinental nuclear missile—is just like a big dick, firing out of its silo, and rapidly flying overhead until it hits its target, where it creates a huge explosion? You think a woman would've invented that?"

Nick: "No, I suppose a woman would've invented something like a nuclear Venus fly-trap, right? Or a milky-colored laser, shooting out of something that looks like an enormous breast? This is ridiculous, Lori. Missiles are shaped that way because that's how they can fly through the air. It's called the science of aerodynamics. Why do all these feminists view everything from the perspective of sex and gender? *They* are the ones obsessed with sex, not men."

Elizabeth laughed: "No, no, men couldn't possibly be obsessed with sex, right? Because women obviously have a bigger sex drive than men."

Nick: "I didn't say that. I'm just saying how silly and narrow it is to view war, and what causes war, in terms of sex and gender. It's so much more complex than that."

Lori: "Does it have to be though? And is it so outrageous to say that war is ultimately caused by men? By male aggression and competitiveness? By a man's tendency to cope with his anger and frustration through violence? By, perhaps, men seeking relief from their problems through physical release? Susan Farudi suggests that war is rooted in a myth about males needing to be the protector of the females."

Nick: "It's just such a negative portrayal of men and, by default, such a positive portrayal of women. Men are to blame, isn't that what the feminists are all about? Men are the cause behind all these serious problems, like war, whereas women are these wonderful, angelic beings. As if women don't get angry, don't make mistakes, aren't ignorant, can't be destructive, are never competitive, and so on. All these sweeping generalizations about men and women, when people and situations should be judged individually. And women have been involved in wars. Margaret Thatcher, for example, ordered Britain into the Falkland Islands in 1982 and then the Persian Gulf War in 1991. If women had the power, there would be just as many wars as men have ordered."

Elizabeth: "I doubt that but, even so, that's Lori's point. Feminists view things from the perspective of sex and gender not because they are obsessed with their own breasts, but because they are concerned with power and equality. And historically, in almost every society, men have had more power than women, and women have suffered widespread inequalities. Almost in every sphere of life. And it's not fair. Women encounter this every day. Men don't, they are used to being in charge. So, yes, feminists can sometimes sound a bit bitter, and obsessed, but it's not because of their own obsessions with gender, it's because of these pervasive issues of power and domination. You'd be obsessed, too, if you felt that your whole life, you've been held back by the other gender. And so, viewed that way, war isn't so much about concrete things like penises, testosterone, and the sex drive as it is about yet another way in which men assert and cement their control over the social world, which they use to serve themselves and not to allow women equal access and fair opportunities."

Lori: "Well said. I wouldn't totally dismiss the dick thing, though. For instance, wasn't Napoleon impotent? Couldn't get it up and give it to women the regular way, so he became a warmongering dictator. Worked out his frustrated sex drive through violence and dominating others. Perfect illustration. Freud would be proud."

Nick: "Well, I've often thought of war being about power and dominance. I think that's an important thesis we need to explore. And you're right about Napoleon, Lori; he had real problems getting it up. But did Sigmund Freud actually say war came out of a frustrated sex drive?"

Elizabeth: "Actually, I know a bit about this. He developed a quite interesting theory about what causes war."

Nick: "Please, go ahead."

Elizabeth: "Well, I guess ultimately it will be connected to the frustrated sex drive thing, or war as an expression of sex, or sexual violence. We all know that, for Freud, it must relate to sex. This was the ultimate reduction of human behavior: everything explained in terms of sex, lack of sex, taboos about sex, fantasizing about sex, analogies to sex, and so on."

They all chuckled.

Elizabeth: "But he makes several important points. First, he noted that people are behind all wars, and so the cause of war must be found within people. In particular, within people's psychological make-up. Wars are created first in people's minds. So, we have to look *there* for the causes of, and solutions to, war."

Nick: "Makes sense."

Elizabeth: "Yet Freud, if he were listening to our discussion, would say we are making a huge, unfounded assumption."

Lori: "Which is what?"

Elizabeth: "Which is that war has a graspable, rational cause behind it. War is about power, or about economic resources, or about spiritual values, or about scheming war profiteers, or about male violence and the drive for dominance, and so on. We've been assuming there's some kind of lack-of-thing, or presence of some other thing, and war is like a solution to a problem. We make sense of war by saying it must be an attempted solution to some perceived problem."

Nick: "How else to view it, though?"

Elizabeth: "Well, perhaps as something which defies rational explanation. Maybe war is, fundamentally, an irrational activity. I mean, look at how horribly destructive it is. War is the single most destructive thing humans have yet invented. How can we possibly think that something so destructive is a solution to anything? Or a way to get ahead? Or to grab resources? Or to spread ideas and values? Freud would say it just doesn't add up."

Nick: "Well, maybe."

Elizabeth: "In 1920, I think, Freud wrote a book called *Civilization and Its Discontents*. The timing is very important; 1920, just after the end of the First World War. Many people have noted how tragic, pointless, and utterly devastating that war was: sowing the seeds of so much of the history of the twentieth century and beyond into today, especially here in Europe and also the Middle East. One of the things most commented on, at the time, was how such a terrible war—such a slaughterhouse of destruction, which achieved so little good—could have come out of a period which had, for so long before the war, been peaceful and prosperous, stable and progressive. Freud devoted himself, in this book, to unlocking this conundrum."

Lori: "And what did he say? What's the key to the puzzle?"

Elizabeth: "The key, I guess, is psychology. It's to understand what drives the human mind. And what drives the human mind isn't always rational. It's not always rational calculation designed at getting more, at getting ahead. Sometimes it's nonrational, or even irrational, impulse. Freud argued that we have two basic urges: creation and destruction. Ultimately, I guess, these urges are connected to sexual desire and its frustration. But I forget which comes first for Freud: are creation and destruction primary, and they find expression in sex? Or is sex primary, and creation and destruction become forms of sexual expression? Anyway, I'm not sure that's crucial for our purposes. I think what Freud adds to our debate about the causes of war is the important idea that *war need not have a rational cause*. War might have an irrational cause; in particular, one rooted in a nonrational, basic drive we have for destruction. And that's how he explained what his contemporaries found so baffling: it was no mystery to him how one could go from an era of great peace, security, and prosperity to one of war, chaos, disease, and poverty. History cycles between eras of progress and regress, precisely because within the human mind there are these twin impulses towards creation and destruction."

Lori: "Hey, I just thought of something."

Elizabeth: "What?"

Lori: "My whole kids-on-the-playground analogy. Freud would have a different spin on it than Marx's thing with property. Well, maybe let me use a slightly different analogy: kids on the beach. Some kids love to create; they spend the whole afternoon making a sand castle. Along come the other kids and they just want to smash it."

Nick: "Or even, sometimes, it's the same kid; he builds something, likes it, takes pride in it, but then smashes it all the same."

Lori: "I've done that."

Nick: "Me, too."

Elizabeth: "Ditto. And it really is perplexing why, isn't it? You said it well, Nick. Even if you like the thing you created and it took time, effort, and commitment, and maybe money, too, you can still want to smash it, and you go ahead and do it."

Nick: "It can provide great release and satisfaction. Is that the connection to sex? It's like tension-building and then tension-releasing."

Elizabeth: "I think so. Creation—civilization—takes a lot of focus and effort, energy and care, and resources and concentration. Which all ratchets up the tension. The tension needs release. Sometimes, we do this nicely—through sports, through sex, and so forth—but sometimes it just comes out in the desire to smash and tear down."

Lori: "Is it a cleansing thing? Or a desire to create even while destroying your creation? Capitalists talk about 'creative destruction.' But as an economic process of tearing down the old to make way for the new."

Elizabeth: "Well, I don't know about that. But it's a neat concept. For Freud, it's got nothing to do with economics. And I don't think he'd want to reduce the appetite for destruction to an appetite for creation, so that in the end all human behavior switches over to the one drive for creation. That would be too positive a spin on the facts, for him. Creation is positive, and amazing and delightful. But there's no denying the darker impulse of human nature, and when you look at those dark impulses analytically, you see that what they all share is an impulse to tear down and destroy. And often, it is just that: an impulse; a drive; and not, say, a fear rationally motivated by a threat; or a rational plan to better things for the future. It's more like Nick's kid example: of the boy tearing down his own creation just because he feels like it."

Lori: "What would Freud then say about prospects for solving the problem of war?"

Elizabeth: "Slim to none. He's gloomy. There's no solution to the problem of war unless you can change human psychology. Unless you can figure out how to get the kid not to want to smash his own castle and, more to the point, get other kids not to smash it even before he himself has a chance to do it. Freud views history as a never-ending cycle between periods of creation and periods of destruction. Thus, progress is an illusion. Or, it's real, but only temporary. *Creation will be followed by destruction.* The best that any of us can hope for is to live most of our lives in a place, and in a period, of relative progress, peace, and prosperity."

Nick: "Bleak. But not untrue, certainly if you look at history seriously. I'm not yet willing, though, to give up on some rational cause of war, like the pursuit of power. Can we move on to that?"

Elizabeth: "Sure, I'm done with Freud."

Nick: "See, this is what I believe is behind all wars: lust for power. To get what you want. To make others do what you want. To get more land, more stuff. As much as you can handle. But it's not for the sake of greed. It's for the sake of ego. For feeling in control: over oneself; over others; over the environment and the circumstances you find yourself in. It's about the 'will-to-power.'"

Lori: "That's Nietzsche."

Nick: "Yup. But many have said the same thing. For example, we spoke of Clausewitz when we were back in the room with the doctor. Clausewitz said that 'war is the continuation of policy by other means,' and he drew a metaphor between a war and an old-style aristocratic duel, like a sword fight between two noblemen over some point of honor."

Lori snorted.

Nick: "Well sure it's macho—and perhaps mega-macho—but stuff like that happened, and forms of it still do today. The point is that he said a war is like a duel, and the point of both is to 'inflict your will upon the opponent,' or to 'bend the enemy to do your bidding, and accept your will.' Something like that. It's all about power."

Elizabeth: "Power defined how?"

Nick: "Precisely *as your ability to get what you want. You go to war to try to get what you want, right?* Whether its tension-release, money, resources, land, spreading your values, feeling better about yourself, or what have you. You're going to war because you think it'll get you what you want. Even if it's a defensive war, and you just want to repel the attacker. It's the logical core shared by all those other theories. War is all about power. War is an assertion of power. War is a response to power; and war's objective is power."

Elizabeth: "But do you mean personal power, like Freud? Individual assertion? War as being rooted in human psychology?"

Nick: "Well, yes and no, I guess. I mean, there's no denying Freud's point about war resulting from the decisions, and actions, of individuals. But there's also no denying such decisions and actions take place in a social context. War is a social enterprise."

Lori: "So what are you saying?"

Nick: "I guess I'm saying the key is both psychology *and* sociology. War's about people pursuing power within their own group. But the key, and here I guess this is what my grand conclusion is: *war is about the drive one group has to dominate another group.* To bring the second group under its authority, and make it obey and do what it wants."

Lori: "Sounds pretty primitive to me, Nick. Pretty 'pack-of-apes-like.' And weren't you the one going ballistic when I tried to draw the exact same analogy?"

Nick: "That was over sex, Lori. This is about power."

Lori: "Sex and power aren't connected?"

Nick: "Well, maybe they are. In fact, sure they are. But in a way which fits my thesis entirely. Sex is an expression of power, not the other way around. Sex flows from power. By definition, no? Power is the ability to get what you want. So, if you've got sex, you must've had the ability to get it. So the pursuit of power is prior to that of sex because, without power, you're empty-handed."

Lori: "Well, maybe your own hand is full, as that's all you've got to service yourself."

They all laughed.

Elizabeth continued: "But I see what you're saying, Nick. All the other theories share the ability to get what you want as a common denominator: you can't spread ideas without power; you can't gain territory, or resources, without power; you can't release tension or destroy, without power; you can't defend yourself without power; you can't attack another without power. So power must really be it. The main objective of war. And that would fit Clausewitz's definition of war as the continuation of policy by other means, since policy is all about power, the ability to guide and direct the way of life of a given group in a given land."

Lori: "Hey! Then we're back to my thesis back in the room with the doctor, in connection with defining the essence of war. It's a violent struggle for power between two groups over who gets to control a territory, and who gets to say what's going on in that territory. Looks like a consensus is developing . . ."

Nick: "OK, Lori, hold on. That's good, that's very good, but let's strive to be thorough, since I've thought of a few more things people might say cause war."

Lori: "Shoot."

Nick: "Well, what about the feelings of nationalism and patriotism which you've been stressing and criticizing? If you look at wars like the First World War—which we've mentioned—you see how this love-of-group feeling, this culture of identity with, and belonging to, a group, was absolutely central. And maybe these feelings can be tied into Freud's account, in that they aren't rational. I mean, are they? Nationalism and patriotism are usually seen as matters of heart, not of head."

Lori: "Marx thought so. He thought nationalism and patriotism were stupid, sentimental emotions designed by the ruling classes to make the working classes happier with society and less likely to revolt. Things like flags, national anthems, and the worship of national heroes are all symbols deliberately structured to take your mind off your economic self-interest—the only thing that's really real—and to incline you to make mushy-headed decisions about loyalty to your country. Or even to go and fight and die for your country." Her cheeks flushed, and tears came to her eyes.

Nick: "Now, now, Lori. Let's not criticize Gil indirectly. He needs our support right now."

Lori flashed: "Of course he needs our support right now! I'm not denying that, am I? I'm his wife after all! I have more at stake with his future than you do! But, if anything, this just gives me more of a right to comment on Gil's bad choices. Where did his love of country get him? All shot up, maybe on the verge of dying. Was that in his interest, or mine? And who benefits from Gil's sacrifice? Not me. Not him. Comfortable Americans back home, who get the increased peace and security from Gil's efforts over here, that's who. You see what I'm saying, right? There's a connection between the nationalism stuff and the earlier military-industrial complex stuff: the lower classes, out of which many soldiers and sailors, marines and airmen come, get into the military out of patriotism. Yet their patriotism gets them in trouble, and the beneficiaries of their sacrifices are those well-positioned people back home, who notoriously don't fight themselves. Guys like Gil are just pawns in a high-stakes chess game played by the captains of industry and the top politicians. The upper classes urge the lower classes, and certainly the soldiers, to be as patriotic as possible, as it makes them more pliable and useful."

Nick: "Well, now you're tarring me with the same brush, and I have to respond. I don't deny that I love America. I consider myself a patriot. Maybe other people benefit from my patriotic motives in joining, and staying in, the military. In fact, I hope they do. But I do it for my own reasons, too. It makes me happy. Well, that's not quite the right phrase. It's more like I can't imagine myself not being in the service, not being a proud and useful American. It's part of who I am."

Elizabeth: "And does that have to be irrational? You were saying, General, that you suspected nationalism and patriotism were nonrational feelings. Lori pounced on that. But it seems to me that it is totally rational to stand up for a part of yourself. It's quite sensible to want to defend and preserve part of your own self-identity."

Nick: "Thanks for that, Elizabeth. I was having a moment of doubt, I suppose. Thanks for getting me back on track."

Lori: "No, Nick, you shouldn't thank her: you should instead go back to your moment of doubt. Because where did this patriotic feeling first come from? It's not a natural part of yourself; rather, it's been programmed into you by every social institution in sight since the very moment you were born. And who controls social institutions? It's not the poorest-of-the-poor, I'll tell you that. Marx was right; patriotism is for soft-minded suckers. It's the political equivalent of thinking with your dick."

Nick: "No, it's not. You're letting your anger cloud your judgment. Elizabeth is right; my self-identity includes being an American. That's nothing for which I ought to apologize. I grant you, Lori, that sometimes people might let their patriotism affect their rational self-interest. Like when sports fans go nuts over their favorite team, or when people say stupid things like 'My country right or wrong.' I love my country, but only want to support it when it does right."

Lori: "So you're willing to go to war just to belong? Just to preserve the part of you which identifies with membership in a group? That's pathetic, Nick. So you're saying the cause of war is group loyalty? Tribal identification? The need to fit in? Like, what is that? Peer pressure?! The desire to fit in? This justifies warfare?! I can't believe what I'm hearing."

Nick: "I'm not necessarily saying such feelings *justify* war. That's an ethical concept. Maybe we can talk about that shortly. All I'm saying, right now, is that such feelings seem to be part of the mix which is present in *explaining* the outbreak of war. Explaining it as a fact, not justifying it. And how is that controversial? How can you deny that? You're just upset about it because it plays into Gil's situation. Clearly, though, group loyalty, self-identity, the desire to belong all play into what causes war."

Elizabeth: "I guess the question, given your remarks about the power thesis, would be whether such things are part of the pursuit of power. Or is something new added by the patriotism thesis?"

Nick: "Well, I'm not sure. Maybe that's why I brought it up. War is caused by the pursuit of power. Maybe the sense of power includes the assertion of one's sense of self, includes the potency of one's group membership, and what one is willing to do on behalf of one's group."

Lori: "Personally, I think this is morally revolting. You can't justify participating in war because it is a means for you to feel like an accepted part of the group. That's like saying it's alright for the members of an urban street gang to beat up and mug innocent bystanders because it gives them a sense of power and belonging. It cements them as a group. No way."

Nick: "But just watch it with the preaching, Lori. I didn't say the nationalist, or group-belonging, motives made it right: merely that the motives play into *the explanation of why war breaks out*. War is between groups of people, right? By definition. So I submit the group dynamic, and the desire to fit in, must play into it. That's all."

Elizabeth: "I'm still wondering about the fit with the power thesis. I think John Keegan had a way of describing what you're saying, in his great book *A History of Warfare*. He pays particular attention to the war which destroyed Easter Island."

Lori: "What's the relevance of that?"

Elizabeth: "Well, Keegan's a Brit. Top-notch military guy. Historian at Sandhurst, which is England's West Point. Great books on the two world wars. He's thought very deeply about what causes war. He suggests it is a sort of primitive, tribal striving for supremacy."

Lori: "The power thesis."

Elizabeth: "But it's more the group version of the power thesis, so to speak. The anthropological thesis, if you will. War is produced by group competition and tribal rivalry. War is a struggle to see which group will call the shots in a given territory."

Lori: "We've covered this. What does Keegan add?"

Elizabeth: "Uh, his huge authority as an expert on war. And this Easter Island example. The groups on Easter Island went to war and ended up destroying the place as an environment fit for humans to live in. They cut down all the trees, slaughtered each other, destroyed each other's crops and children, and the only thing left today are the huge, creepy, stone face statues."

Nick: "Why does he think this happened?"

Elizabeth: "Precisely because it grew from small scale, rationally graspable issues—like the struggle for political power and more economic resources—to huge, deep, and quite nonrational drives, like the need for one's group to win regardless of the cost: indeed, even if the cost was doing things which would ruin the place for one's own group to live."

Lori: "Freud would like that, it was irrational destruction."

Nick: "But he wouldn't like it, as it has nothing to do with sex. It's all about one's membership in a group, and that group needing to feel better than the group next door. Did they really destroy the whole island?"

Elizabeth: "Well, obviously, else we'd have people on that island. But their group hatreds ran so deep, and the drive for group primacy ran so strong, that they all killed each other. The war became an act of suicide."

Lori: "But then it's about winning, no? Or survival?"

Elizabeth: "Well, I guess war sometimes really is about survival if you're talking about going to war to defend yourself against an aggressive invasion by a brutal attacker. Like the Polish against the Nazis in 1939. But someone like Keegan would say that, if you look carefully at history, only very rarely is genuine survival at stake, whether for people or individuals. Politicians just talk that up to heighten the sense of the stakes and motivate their citizens. But the facts show, overwhelmingly, it's not really about survival but, rather, about a group having its say prevail in a given territory."

Lori: "And the drive to win?"

Elizabeth: "Well, is that different from the power thesis? You want to win, which is to say, you want your will to prevail. You want to be able to impose the terms of the peace."

Nick: "Even if 'the peace' is the peace of the grave? The destruction of everyone around, like on Easter Island?"

Elizabeth: "Well, there's certainly such a thing as a Pyrrhic victory. Named after a Greek general who beat the Romans in a battle, but the casualties were so enormous the victory was beside the point, and the Romans eventually conquered the Greeks anyways. Or maybe even more, you'll actually lose the basis for a functioning society, as on Easter Island. War as suicide."

Nick: "Irrational. Better to die oneself than to lose, or to admit loss, or to admit being a member of a loser culture or a destroyed society."

Elizabeth: "Well, it's only irrational if you fail to see how the drive to win and to secure power for one's group takes on a dynamic all its own, almost like a gambling addiction. People and societies caught up in the dynamic can't step back and view it objectively. In any event, whether rational or not, the drive for group primacy and power, I would think, have been established as major—perhaps THE—motivating causes of war."

Lori: "Do we have to reduce it to one cause though? That's so reductive: attempting to distill all the various causes into one common denominator. Why can't we rest with the common-sense position that various things cause various wars? Power, greed, group rivalry, and patriotism: they can all play a motivating role, and sometimes one factor, or group of factors, plays a more obvious role than the others."

Elizabeth: "Well, that's a classic clash between monism and pluralism: between 'mono-causality' the search for, or insistence upon, one cause—and 'many-causality,' noting how diverse or plural causes are playing a role. I guess I favor monism in two circumstances: 1) where it seems accurate, that is, where it really seems as though there is one shared common denominator; or 2) because it is more insightful and simple. You don't have to sample all the various causes, and describe and test them all against the evidence. You can just focus on the one, and it gives a fully satisfying explanation. But crucially, it must offer a complete and insightful explanation: you can't just pick one cause because you're lazy and like to keep it simple for its own sake."

Nick: "And I really think the power thesis fulfills those two conditions: it certainly is more simple than constructing some complex cause of war—a calculus which would weigh the various factors we've considered. Next, it really does seem to be the logical core of what we're getting at, and what we've been investigating at such length, here in the mess hall. Get it: *mess* hall? Ha! The power thesis cleans up the mess and puts the cause of war squarely into the right slot: *people, and groups, go to war because they are seeking power over others.* Sometimes this power-seeking is rational; sometimes not."

Lori: "Well, I'm not prepared to admit defeat. I don't see the problem, or the trouble, with settling for the many-cause thesis. But, even if I admit the striving for power is very important, what hope does such a thesis give us, in terms of a solution for war? What's the solution to humans seeking power over others?"

Nick: "There is none. That's why war has always been with us, and why it always will be. The will-to-power is part of human nature, and we can't change that. Only God can. Our sinful nature contains the lust for power, domination, and supremacy, and there's no cure for it on this side of the grave. Only God's redemptive grace, in the next life, can cleanse us of our supremacy-seeking."

Elizabeth: "Well, I don't share your religious outlook, General. But I do share the thinking that, since war has always been present in human history, its main cause must be pretty basic and very difficult to get rid of or to change. Otherwise, we would've done just that, right? And clearly, looking around, people love trying to control other people. They love telling other people what to do. They love feeling superior to others, and love trying to get other people to do as they do. Until we can curb or change that, it's hard to see a plausible solution to war."

Lori: "But wouldn't such a solution be *both* to make people feel more empowered, so they are less likely to have that drive, *plus* condition people not to seek domination over others? Drill it into children's heads over and over, and wait a few generations for the effects to be felt. I'm not impressed by arguments regarding human nature, much less God. People behave the way they do because of the environment they find themselves in. And the human environment is dominated by human institutions: change the institutions, and you change the people. If people felt more empowered themselves, and were trained to feel shame at trying to dominate others, then we could curb the lust for power and thus the incidences of war."

Nick: "Your beliefs, Lori, are a bizarre blend of paranoid, pessimistic conspiracy theory, and wildly optimistic reformation."

Lori: "I love you, too, Uncle Nick."

Nick: "No, I'm serious. I'm not saying you're inconsistent, because you're not. You think the current situation really stinks, but you believe it can and ought to be changed for the better. I like that aspect of hope you have."

Lori: "Thank you."

Nick: "I just don't think you can change human nature. All the institutional change in the world won't make a human 'un-human.' War is caused by the drive for power, which is part of human nature. New laws, new schools, new governments, they don't change that. So war will always be around. That's just realism regarding war."

Just then, a handsome, young African-American corporal, in full uniform, walked into the room and up to the group. His name was Carter Johnson.

Chapter Four

Power and Pessimism

Carter Johnson: "I'm Corporal Carter Johnson. I've been instructed to give you all a tour of the base, if you wish. Dr. Leung says the latest procedure on Major Gordon will be awhile and there won't be anything to report for some time. Maybe not even until tonight."

Nick: "I'd love a tour! Nice to take a brisk walk after a good meal. Plus, it looks like there's nothing better to do anyway, right? And you know, after all my years in the Army, I've never been on this base until today. It's huge. Ladies, what do you say? Would you like a rest, or would you prefer to join us? How's the weather outside, Corporal?"

Carter: "Nice. Chilly, but with sun."

Elizabeth: "I'm up for it. A nice walk on a nice fall day."

Lori: "Me, too. And Nick, Dr. Leung has your cell number, in case of . . . uh, any sudden developments, right?"

They were all quiet for a moment.

Nick: "Indeed, he does. Let's walk."

As the group moved about the base, Nick continued: "So, as I was saying, realism affords no solution to the problem of war, as it's *all* about power, and the pursuit of power is a part of human nature, and so that's why war has always been, and always will be, part of human history."

Carter: "General, are you talking about realism, as in the doctrine about war?"

Nick: "Yes."

Carter: "Then with respect, Sir, I think that's not true. Or, not the whole truth."

Nick: "Oh really, Corporal? You should join our conversation about war, and shed some light on the issue."

Carter: "Only if I'm not interrupting. I mean, I am supposed to give you a tour, pointing things out, and all."

Elizabeth: "Well, you can do that on the way. Please, join us and tell us about realism."

Lori: "Yes, but first: how do you know about it?"

Carter: "Well, I actually completed my master's thesis on realism and war strategy, just three months ago."

Lori: "Really? I'm writing mine right now, on the American empire. Where did you get yours?"

Carter: "Oklahoma State. Paid for by Uncle Sam."

Nick: "Nice. That's the way to go." He looked at Lori to check the response, then continued: "OK, young man, so you've earned a master's degree. Big deal. Enlighten us on the deeper meaning of realism."

Carter: "Well, my narrow point is just that realism *does* support a view regarding how to solve war. It's not just that we're condemned to have war because war is part of human nature."

Nick: "But we've always had war. How can that fact give anyone confidence about a solution to warfare? I read this great book one time, a while ago now: I forget the plot but it was really good . . ."

They all laughed.

Nick: ". . . anyways the title was *"Bred in the Bones."* I like that expression, and think it speaks to this point—the pursuit of power is bred into our bones. It's part of who we are, as fallen and flawed humans. There's nothing going to wash it out of us, not a more equal distribution of resources, nothing. War is here to stay."

Lori: "Whereas I think there's got to be a way, at least to make it somehow better. Fewer wars. Less destruction. If you change the environment in which people operate, they'll change their behavior."

Carter: "The realists agree. Or, at least, some do. There are actually many realists—different kinds—and some agree with you, and some agree with the General."

Nick: "Call me Nick."

Carter: "Thank you, Sir, I will. Nick's views are supported by people like Reinhold Niebuhr. Sir, am I correct in inferring, from your comment about 'fallen beings,' that you're religious?"

Nick: "Yes, but just for the record that's my own personal view and has nothing to do with the Army."

Carter: "I understand. Have you read Niebuhr?"

Nick: "No."

Carter: "I haven't either. But I hear he gets labeled as a 'Christian Realist.'"

Lori: "What the hell is that?"

They laughed again.

Nick: "Hell, indeed! Who the hell is a Christian Realist?"

Carter: "Someone who draws a sharp split between this world and the next. Their vision of this world is bleak and pessimistic, stressing humanity's flaws and vices, its stupidities and weaknesses. Very much in line with the realist vision you seem to believe in, Nick. The lust for power producing violence and war, and this lust for power being bred into the bones of human nature."

Nick: "Totally."

Lori: "But Nick how is such a picture of human nature consistent with being Christian? Like where's the love and the forgiveness? Where is the redemption?"

Nick: "Well, I guess that comes later. Or it comes through the grace of God."

Carter: "And that's the dualism: the two worldview. I think Augustine gets credit for it originally: there's this world—the City of Man—and it is dominated by sin, lust, weakness, vice, and conflict. And thus war. But then there's the next world—the City of God—and it is governed by love, forgiveness, and redemption, and hence peace."

Lori: "So, on this side of the grave—in the City of Man—there will always be war?"

Carter: "Yes."

Lori: "And our only hope of escape is a religious hope? God is the only solution to the problem of war?"

Both Elizabeth and Lori laughed, but neither Nick nor Carter did.

Lori pressed her point: "But, as I recall, God in the *Old Testament* is pretty warlike. Doesn't God actually order the Israelites to slaughter the Canaanites?"

They all shrugged their shoulders. No one knew for sure.

Nick: "But the *Old Testament* is incomplete without the *New Testament*, and there we have the example of Jesus' love and redemption."

Elizabeth: "And peaceful ways."

Nick: "And the humans—the violent humans—at the time couldn't understand his peaceful and loving ways, so they killed him. They were afraid of letting go of their normal, weak, and fearful values. See: violence is part of human nature."

Lori: "But it's just giving up way too easily on the problem of war. 'We're hopeless; please send some divine intervention, as that's our only hope.'"

Carter: "Well, maybe that's where the other kind of realist comes in—the more modern kind. The kind which views war not as a problem in human nature, but rather, as a problem about the structure of the world. Well, not so much the structure as *the lack of structure*, in the world."

Elizabeth: "Say more."

Carter: "Well, the modern realists—not Christian Realists like Niebuhr—they recommend realism as a strategy of choice, not as something to view as coming out of our bones."

Lori: "Yes, of course. It's a choice, not a necessity."

Carter: "Right, war is a foreign policy choice. An option a country has, as it decides how to behave on the international stage. Do you go to war, or don't you? What's the smart thing to do? Anyway, in foreign policy, there's this classic distinction between realism and idealism. With realism, you essentially adopt a selfish strategy of maximizing your nation's best interests."

Elizabeth: "And how are 'interests' defined?"

Carter: "Well, it's all about power, really. The ability to get what you want. That's the ultimate thing to be interested in, right, the general ability to get whatever else you might want? But realists typically distinguish between hard and soft power."

Nick: "Here we go with the feminism again. Get it? Hard versus soft."

Lori: "Yes, Nick, brilliant. Cool off and let Carter continue."

Carter: "Hard power is composed of two things: the military and the economy. Essentially, bullets and bucks. Guns and money. These are the most potent and immediate ways of forcing the world, and others in it, to give you what you want: threaten them with violence; actually force them with violence; or bribe them with cash incentives."

Elizabeth: "And soft power?"

Carter: "Soft power deals with cultural impact, and getting others to agree with your worldview. So things like media, religion, education, and political and moral values. Even, I suppose, things like art, or music, or Internet content."

Lori: "This revisits some things we've discussed."

Carter: "Oh, I'm sorry. Shall I stop?"

Lori: "No, no, please don't. Just noting how some things are coming together. So, for the realist, it's all about selfishly getting as much hard and soft power for your country as you can. That's your country's top interest in today's world."

Carter: "Absolutely."

Elizabeth: "Now, why is that the top interest?"

Carter: "Here's where the strategy comes into it. I mean, Christian Realists would say this is a country's top interest essentially because there's no other choice, it's who we are, bred into our bones. But Strategy Realists—let's call them—say: 'Look, we can put all controversial, speculative claims about human nature and God aside and just say that sticking to the pursuit of power is the most *rational* foreign policy a country can make.'"

Elizabeth: "And how's that?"

Carter: "Well, here's where the contrast with idealism comes in. Whereas realism is about selfishly trying to do the best you can for your own country— 'Looking out for Number One'—idealism counsels instead trying to do whatever your country can to make the entire world a better place. It's more of an altruistic policy involving a giving of yourself to the world. Contributing what you can, to improve everyone's lot."

Nick: "Selfishness versus altruism."

Carter: "Yes, basically. But from the perspective of countries instead of individuals. This is how Strategy Realists view it: your country has a very basic choice to make about how it wishes to interact with other countries. Your country can do so selfishly, or it can do so altruistically."

Elizabeth: "So, why choose selfishness over altruism?"

Lori: "Yeah, why not follow your ideals and try to improve the world?"

Carter: "Essentially because it is not the smart thing to do. Here's why, and finally we're at the point where I can say what I meant about 'the lack of structure in the world.' The key thing for Strategy Realists is that there is no effective world government. If there were, we might have our solution to the problem of war."

Nick: "Wait, *world government is the solution to the problem of war?*"

Carter: "It might be. But I need to come back to that in a minute."

Nick: "OK, interesting. Please continue."

Carter: "For Strategy Realists, it's all about behaving in a smart and rational way, considering the environment in which you find yourself."

Lori: "And idealism is dumb? Altruism is stupid?"

Carter: "Some Strategy Realists will come out and say that, but most prefer to say that idealism is not the best bet. There may be some rationale behind sticking to your ideals and wanting to make the world a better place but, when push comes to shove, you have to stick to protecting your interests."

Nick: "Truer words were never spoken."

Carter: "What motivates it all is precisely a focus on the circumstances or the environment, as opposed to Christian Realists, who focus on the nature of the players or actors *within* their environment. The starting point is that there is no effective world government."

Lori: "There's the United Nations."

Nick: "But c'mon, Lori, it's no world government!"

Carter: "That's right. It's not a government the way a well-run national government is. A well-run national government provides, in general, law and order for a society."

Lori: "No, it doesn't. What about criminals? Even in America they have criminals."

Carter: "A well-run government doesn't mean a perfect government. There's no way to predict and prevent every single possible criminal action. The point is that, generally, there's peace and order and, when bad things occasionally break out, the government has a whole series of people and procedures in place to put things right. You know, like cops and the military, and then the court system."

Lori: "OK."

Carter: "And a well-run government is effective and legitimate, it generally achieves its goals, has the tools to do so (like tax dollars), and the people it rules over by and large accept its leadership role in ordering their society."

Nick: "So you can see how the UN does not satisfy those conditions: it can't collect taxes; it often is ineffective; and certainly it does not keep the peace, or international law-and-order. It is a voluntary association between countries, not a government ruling over them."

Carter: "Precisely. So the reality of the international system, as realists put it, is that it is an ungoverned condition. Some say it's a flat-out anarchy, a global free-for-all like the lawless 'wild, wild west.' Others say there are pockets of order, as secured by powerful countries . . ."

Lori: "Like the American empire!"

Carter: ". . . yes, that's right. But that order is only partial, not whole. It's only regional and not global. Plus, it's always being challenged, and could be undermined by rivals at any time."

Elizabeth: "OK, so that's the nature of the environment we find ourselves in: there's no world government, and what's left is only partial international order, at best, or total global anarchy at worst."

Carter: "Right."

Elizabeth: "So how does that make selfishness rational?"

Carter: "Because the environment is fundamentally insecure, and you can't assume that the other players won't be selfish themselves."

Elizabeth: "Why can't you assume that?"

Carter: "Because information is imperfect, data is not complete, people hide their true intentions, and in the case of countries, there are the barriers of language, culture, and historical development."

Lori: "This sounds like a game of chess. Or 'Risk.'"

Carter: "In many ways, yes. Everyone is each trying to do the best for themselves; there is no one ruler, only players; and you don't know the full real strategy of the other players. Is it rational to trust the other, and be altruistic in such circumstances? No, at the least, you play to protect yourself; at most, you play to win. So you know where this is going, there's insecurity, because there's no world government. There is no general peace, and you certainly can't count on any global police, or world court procedures, coming to your rescue in case things go wrong."

Nick: "You can say that again. In the final analysis, self-help is the order of the day. You must see to it yourself and for your people."

Elizabeth: "You can only count on yourself."

Lori: "And you can't, in the end, trust anyone but yourself. Because you know so little about them, the others. Or, what you *do* know shows that they will behave selfishly, and so you should, too."

Carter: "You can't get suckered. This is a key concept for realism: behaving in an altruistic way exposes you very seriously to being exploited, and taken advantage of, by others. But that's not rational behavior. What is rational is doing the best you can for yourself. And, given the nature of the game—so to speak—that means focusing on power. Getting as much hard and soft power for your society as you can. The bottom line on global strategy for realists is this: at the very least, protect what you've presently got—don't let things get any worse. At most, though—if it's possible for you—strive to remake the world in your own image."

Lori: "And that's what America is trying to do right now with its empire . . ."

Nick: ". . . if it *had* an empire . . ."

Lori: ". . . and then it gets so shocked when other countries resist, or tell America they don't especially approve of American domination."

Carter: "Realism does tend to be a very influential doctrine among the elite decision-makers in U.S. foreign policy. Has been since WWII when America became one of the top powers. Henry Kissinger, Secretary of State under President Nixon, was a prime exponent of the doctrine, uh, I think in the '70s."

Nick: "But then what's realism's attitude towards war? How will world government solve war?"

Carter: "Two very big questions. Let's take the second, first. Since the basic reason for war and international conflict is the fundamental insecurity felt between players, the solution is to create an effective, reliable institution which can regulate the players and, above all, become an institution that anyone can reliably turn to for the solution of some grievance. Like in domestic society; when we've got a real problem with some individual, we don't reach for a knife or a gun. Well, dumb or drunk people might do that, but reasonable people call the cops, and/or their lawyer, and they begin a process they can count on to address their grievance. Violence in the use of problem-solving is thereby prevented. Draw an analogy to international affairs; if we had a neutral, well-run, well-resourced, effective world government, then countries would turn to it to solve their disputes through reliable procedures. They wouldn't have to resort to force."

Elizabeth: "The world government itself might, though—just like the government has to sometimes use cops, or even the military, to put the lid on irrational criminal types, or on gangs and such."

Carter: "Yes, but then I guess by definition it wouldn't be war but rather a kind of global police action or law enforcement procedure. But note that that is *not* just shifting names over the phenomenon of war. There may be no such thing as a perfect world wherein there is never any violence at all—not even the occasional violence of law enforcement—but it would still clearly, drastically reduce the outbreak of wars, and the savagery with which they are fought, if a neutral and effective world government could solve problems between countries."

Nick: "Not bad. Not bad. And so, the realists—or, these realists—would say that the reason why there has always been war, historically, is not because of human nature, but rather, because there has never been an effective world government."

Carter: "Precisely."

Nick: "Hmmm."

Lori: "See, *Nick:* if you change the environment; and if you change the institutions; you can change people's behavior. No one knows what the hell human nature really is, or if it is there or real at all. Ditto for God. I mean, c'mon, who is authorized to claim they know—I mean, really *know*—such things? They are just pretending to know, hoping their act will win them some social influence.

But what we really *do* know, and can grasp as obvious fact, is that here we are interacting with each other directly and through the mediation of very powerful institutions. These institutions—government, the economy, the law, the military, schools, hospitals—have an enormous impact on our lives. Change them and you can change people's lives."

Nick: "Maybe. But how on earth are we supposed to create such a world government—much less one that's effective, reliable, and well-resourced?"

Elizabeth: "Right. It's one thing to have a *theoretical* solution to the problem of war, quite another to see it implemented *in practice*."

Carter: "Well, that I'm not sure of that, and I'm not sure anyone has the answer. That hasn't been part of my own research."

Nick: "Fair enough."

Carter: "But what I will say is that individual countries managed to find similar solutions when establishing their own basic constitutional and government structures—just think of the founding of America and the eventual creation of a new federal government—and so why can't the same sort of thing happen globally?"

Elizabeth: "Because it is so much more difficult!"

Carter: "Would it be, though? Would it *have* to be? I mean, I suppose to an extent, it would, since there are so many more people, more cultures and languages, and a much bigger territory to govern. But that just makes it more difficult *in quantity*, right? Not *quality*. It is still the same sort, or same set, of issues to be resolved: how to set up the government; who gets to be an official; procedures for elections, and checks-and-balances; on which values and principles would the government be based; and so on."

Lori: "Look, Carter, I'd rather like to agree with you and, you know, I still do. But when you look at how disappointing the UN can be—especially when it comes to war and peace—can we have real confidence that an effective world government could ever become a practical reality? For example, America would never let it happen."

Nick: "Why on earth not?"

Lori: "Because of the American empire. America is the most powerful country in the world. You think the U.S. government would tolerate yet another level of government *above itself*, and to which it would be forced to answer and to *defer* on major issues? Ha! Not in my lifetime."

Elizabeth: "But maybe if America could be persuaded that it would be in its own rational self-interest to support such a world government, then it would do so."

Lori: "But how would that argument work? I just gave you an argument that America's power would be reduced if a world government came into being and, since realistic self-interest is all about power, how would America ever condone that?"

Elizabeth: "Well, I think you may be looking at it in too conceptual a way. If you focus on certain practical problems, like reconstruction in Afghanistan and Iraq, or dealing with global threats like terrorism and rogue regimes, you can see how—if it thought through the issue clearly—the United States would love to have, and would clearly benefit from having, global coordination and substantial help. A good and decent world government could lessen the burden on America's shoulders when it came to global problem-solving. America wouldn't always be the one from whom everyone was demanding action, upon whom everyone was putting responsibility. Other countries could also be required to throw in more cash than they typically do: the United States is easily the biggest budget contributor to the UN. Moreover, if the United States could use its influence to shape the world government to respect some key American or Western values, then that could end up being quite an effective way of spreading its ideals around the world."

Lori: "That's a pretty good argument, Liz!"

Elizabeth: "Thanks, I try. We lawyers love to lay out a case."

Nick: "OK, Carter. Good on the world government thing. Thought provoking. But now you've got to go back to my other questions: before an effective global government can provide the solution to the problem of war, what does realism recommend we do in the meantime? I mean, in connection with war. When should we go to war? How should we behave during and after war?"

Carter: "It's all national self-interest. Self-interest in national security and the power of a country's foreign policy. So, war should only be fought when it is clearly and manifestly in a nation's self-interest to do so. This is how today's realists generally see it. A while ago, many realists advocated 'a balance of power.'"

Elizabeth: "What was that?"

Nick: "I know this one. The doctrine of the balance of power first evolved in Europe. It was thought that, to keep general peace in Europe, the major powers— England, France, Germany, Austria-Hungary, Russia—had to be more or less equal in power. Or, at least, form alliances which added up to being more or less equal in power. The idea was that a careful equilibrium was the key to peace, with no one power being able to dominate or conquer the others. I think their concern, and memory, was of Napoleon and France, who were so much more powerful than the others, and used their strength to try to aggressively conquer much of Europe."

Elizabeth: "OK."

Nick: "There's more. That was the theory. The practice was that, contrary to the theory's intent, it probably produced more wars. Whenever anyone felt that someone else was becoming too big or powerful, they'd essentially launch a preemptive strike to take 'em down a peg or two. Put 'em in their place—through war! So, the irony was that a doctrine designed to keep the 'long-term peace' actually sparked a bunch of short-term wars. And, since those short-term wars kept happening and happening, where was that 'long-term peace' the thinkers promised? Ha!"

Carter: "Right. Absolutely right. And that's why very few realist thinkers talk anymore about a balance of power. When they write about war, they tend to be very cautious in their advice, as there's all this evidence about how risky and costly war is; and how it can backfire; and how it creates unforeseen effects; and how it can draw in other players; and just generally how it routinely turns out to be much more expensive, difficult, and tricky than anyone expects at the start. As a result, the realists tend to suggest that it is only under very rare, and completely clear-cut, circumstances that today's countries should think about going to war. Like, when your country is attacked by another. You have to respond to defend yourself, to prevent yourself from getting conquered, and to let all the other countries around know that you won't just lay down and die—you'll fight back—and so they shouldn't go getting any ideas, either. Almost everyone agrees that this kind of self-defense is a solid and rational reason for going to war."

No one mentioned any disagreement.

Carter: "It's all the other kinds of war and armed conflict that realists disagree over, and will offer different kinds of advice about. Because, even though they will all generally agree about the centrality of power, and the need for a country to pursue, rationally, its own self-interest, they can and do differ on what the facts suggest about this particular case, or this particular war."

Nick: "Like Iraq?"

Carter: "Absolutely, most realists agreed that the Afghan war had to be done, and was a good idea. America was attacked on 9/11, and so that regime had to be overthrown with force and then reconstructed into something better, namely, a new regime type which would not foster or support terrorism."

Nick: "Absolutely. Should we have just stood there, and done nothing? Then it would've been open season on America, on the part of any nut job, fanatic outfit!"

Again, no one disagreed.

Carter: "But then Iraq was different. Was it in America's rational self-interest to attack Iraq in 2003, overthrowing Saddam's regime? Saddam had not attacked us, and so it was quite different from the al-Qaeda/Afghanistan case. The U.S. decision-makers feared Saddam would attack, as we know, and they were very concerned about the kinds of weapons he might use in such an attack, but the matter of fact remains that Saddam did not attack us before we attacked him."

Nick: "Some people linked Saddam to al-Qaeda."

Elizabeth: "But the 9/11 *Congressional Report* said there was no rational basis for that connection. Saddam was a secular dictator and he despised Islamic fanatics, and the feeling was mutual. It makes no sense for them to become allies."

Nick: "Unless they were allied for the sole purpose of striking America."

Elizabeth: "But the 9/11 *Report* showed no factual basis for believing in such a one-time conspiracy between people who otherwise hated each other and could hardly have been more different."

Carter: "Realists debated all these issues. The great thing about the doctrine is its simple, one-minded focus on rational self-interest. This is the be-all and end-all. One Very Simple Principle, to borrow a phrase. The problems arise, though, precisely when the principle gets applied to a concrete case, like the Iraq invasion. For instance, let's ask ourselves the arguments which a realist could've made to go to Iraq."

Nick: "OK. Get rid of Saddam, most obviously. A dictator to his own people. And a pain in America's ass since he invaded Kuwait in 1990 and sparked the first Iraq War—the Persian Gulf War—in 1991."

Lori: "Plus, secure the oil. Let's not forget the real reason."

Elizabeth: "People also said it would be good to create a real Arab democracy, for the strategic long-term reasons of having to transform the Middle East into a more friendly and pro-Western kind of place."

Nick: "And, look, whether Saddam and al-Qaeda actually did have any links, or whether Saddam actually did have any weapons of mass destruction, President Bush couldn't rationally have taken the chance that they didn't, right?"

Carter: "And that's an important point. For the realists, there's a disagreement about self-interest and risk. Or, more precisely, about rationality and risk. Some realists viewed the risk of *not* attacking as too big—fearing another 9/11—and so they supported the strike and the regime-change war. Others felt the risk of attacking was too great—the war will be risky and costly, and what do we do with Iraq after we win?—and so they advised against it."

Nick: "Seems more prudent to have attacked to me and, look, not one more 9/11 attack has happened since. You know what they say in football: the best defense is a good offence."

Lori: "You're kidding, right, Nick? There was no Saddam/al-Qaeda link. There were no weapons of mass destruction. The war was sold to the public on the basis of a false fear of another 9/11. Iraq is still a hopeless mess and our military men and women—my poor Gil and others—are getting shot up and dying in a failed war. A failed war, Nick! No realist can say it's rational to fight a war and lose!"

Nick: "But we didn't lose! We haven't. We successfully got rid of a dictator! We've so far prevented another 9/11! Those are huge victories. Victories, Lori! Gil has not been injured in a losing cause, damn it! It's just that the reconstruction of Iraq is proving very, very difficult. The war was a success; the postwar is proving much harder to win."

Lori: "Postwar! Postwar?! *Post*?!! Was my Gil blown up in an after-war event— a post-violence bit of violence?! The war is still ongoing Nick, and I don't think we're winning it at all. We're still there, after all these years, and Iraq is not substantially better off than it was prior to that original attack."

Nick: "B.S., Lori. Utter B.S. It's not a dictatorship anymore, Lori. That must, just logically, be better."

Lori: "I don't think so. At least there was order under Saddam. I mean, he was a rights-violating dictator. But he kept the peace in Iraq. Now, you have all the groups fighting each other for influence in postwar Iraq. And America can't make substantial headway in keeping Iraq unified and moving forward. We're just there to prevent the whole country from collapsing completely, and some days it looks like we can do this whereas most days it looks like we can't."

Nick: "I disagree, Lori. But, look, I fear we've shaken Carter off his path. Forgive us, Carter. We both have strong views on this topic, and my nephew— Lori's husband—lays injured here on the base because of the Iraq War."

Carter: "Understood. Would you like something more from me?"

Elizabeth: "I would. Throughout, we've spoken of rational self-interest and prudence; being smart in terms of not getting suckered, and advancing one's own self-interest."

Carter: "Right."

Elizabeth: "So, where's the scope for morality? Or justice?"

Carter: "Sorry?"

Elizabeth: "Well, are ideals and moral values really so dumb and stupid that it makes no sense whatsoever to appeal to them, or at least to consider them, when considering these things?"

Carter: "Realists would all say 'yes,' strange as it sounds. Or, wait, I guess some of them might say you ought to appeal to moral values if it will also serve your interests in terms of winning the war. And I guess it usually is in one's interests to do so: that is to persuade one's people, and the international community, that it's a good war to fight and so on. But the appeal to morality is strictly strategic: morality, or common moral belief, is useful to securing the ends of the war in particular and the country's abiding national interests in general."

Nick: "The strategic use of morality, in the service of national interests."

Lori: "But that's just rhetoric. That's not the sincere use of morality. That's not what Elizabeth means."

Carter: "I know that's true. I'm just pointing out how realists tend to view it. Actually, they tend to be divided on this issue."

Elizabeth: "How so?"

Carter: "Well, uh, let me see. I think there's a big split between them here. There are descriptive realists, and prescriptive realists."

Nick: "Huh?"

Carter: "Descriptive realists are those who believe that morality, ethics, and justice in no way influence a state's foreign policy. I mean, a state may lie, or pay lip service to ideals of ethics or justice—again, for popular support, usually—but that is, as Lori says, simply rhetoric. Just making strategic and self-serving use of moral values. A descriptive realist would assert that national governments are in no real way motivated by concerns over morality and justice."

Elizabeth: "So there's no such thing as morality, or justice, in war?"

Carter: "No, not for the descriptive realist. War is utterly amoral, or even immoral. War should only be thought of in a strategic way; will fighting this war serve our interests? What do we need to do to win? For the prescriptive realist, it's different. They think, interestingly, that the need for a state to be motivated strategically on the international stage—including during wartime—is ultimately rooted in a basic moral duty which the government of any country owes to its people, namely, to protect and serve their interests to the best of its ability."

Nick: "OK, OK, wait. We've got two different propositions on the table. They're quite different. Let's examine them with more care."

Lori: "Well, well, interested in details now, huh, Nick?"

Nick: "What? That's not fair, Lori. You're still just mad at me from before. Of course I'm interested in details, especially about realism, which is a doctrine to which I'm attracted. So, *Carter:* descriptive versus prescriptive realism."

Carter: "Descriptive realism says morality and war have nothing to do with each other and never have. War is about as completely amoral—or even immoral—an activity as you can get. It's all about selfish survival and the pursuit of power. What could be more obvious than that? 'War is hell,' as somebody said. Indeed, descriptive realists view all foreign policy choices in terms of pursuing strategic or prudential interests, and any appeals to morality as being either hypocrisy, stupidity, or a piece of strategy—rhetorical strategy to persuade people and garner support for the war. Whereas prescriptive realists view morality as real — not just rhetorical — and indeed as justifying the pursuit of self-interest internationally."

Lori: "Wait, morality requires strategic selfishness?"

Carter: "Exactly, for the prescriptive realist. The most fundamental duty — and it's a moral duty — which a government has is to protect the interests of the people over whom it governs. And physical security against foreign attack would rank very close to the top when it comes to protecting people's interests. So, yes, it's a *moral* duty to behave in an *amoral,* self-regarding way. But — and I stress this — it's not because of the nature of morality, but rather, because of the nature of the arena in which the players find themselves."

Lori: "There's no peace or authority, nor trust nor perfect knowledge . . ."

Elizabeth: ". . . and so a government must — morally must — behave in a kind of nasty, ruthless way internationally, in order to effectively protect its own national population from suffering harm."

Nick: "You know, I was once talking to a guy who said — I think this was back during Watergate, with the huge cloud over the heads of Nixon and Kissinger and the ethics of their conduct — that he actually thought it was better for people like that to be in charge of our foreign policy. 'To defend us well against sons of bitches,' he said, 'we need to have our own sons of bitches.' I thought it was appallingly unprincipled at the time, but now I see some sense to it. It's a rough-and-tumble world. Leaders must be prepared to deal with tough enemies in their own tough way."

Lori: "OK. There's clearly some sense to realism. But we can't have this turn into a love-in, celebrating the greatness of realism. Clearly, many countries behave selfishly. Clearly, you've got to watch out for that. And this fact gives you incentive to behave like a realist, in turn. I don't deny it. That's why it's called realism, I suppose, no? Interesting self-labeling, no? But let's give the doctrine the same kind of scrutiny we gave the other thesis, right? It's only good if it can stand the test of substantial challenge."

Elizabeth: "Exactly. OK, let's take on 'descriptive realism' first. I'm more impressed with the second doctrine, so I'll focus on the first. It seems to me, as an international lawyer, to be clearly false. It's just not true that states are in no way motivated by moral ideals, that it is all just hypocrisy. Even during wartime."

Nick: "Can you give any examples?"

Elizabeth: "Sure, I've often heard it said that Britain's efforts to outlaw the slave trade, in the nineteenth century, can only be explained sufficiently by appealing to a sincere commitment to moral values. The British people simply would not stand for the continuation of slavery. So they pressed the British government to get behind abolition and press for that internationally. And so that's what Britain did."

Nick: "But there's your realist reason, no? The British government strategically responded to the moral concerns of the British people—not because it was moral, but rather, because it was the concern of the people over whom it governed, and you can't totally ignore the wishes of the people."

Elizabeth: "Maybe, but consider this. Anti-slavery agitation, and the eventual prohibition of the international slave trade, dramatically increased the price of imports into Britain of such basic commodities as cotton, tea, and sugar."

Nick: "So?"

Elizabeth: "The point is that they pursued their moral convictions, and held on to them, even when it hampered their economic interests in cheap commodities like cotton, tea, and sugar. I'm suggesting this is *evidence of a genuinely moral motivation in international affairs*—something which, if true, would disprove descriptive realism, which denies states can be motivated by anything other than prudential self-interest."

Carter: "But wasn't it in the interests of the British government to please its own people?"

Elizabeth: "Maybe generally, but it wasn't a true democracy at the time, so I don't think you can say its decision was motivated solely by a desire to please the people."

Carter: "The other example people give of the sort of thing you're advocating is American involvement in the First World War."

Elizabeth: "That's actually a very good example. Why did America get involved in that war? What national self-interest was vitally served by participating in that exclusively European conflict? It would seem none: it's not as though U.S. territory was invaded, or U.S. business interests were at stake; rather, it was Woodrow Wilson's sincere commitment to the spread of democracy."

Nick: "An idealistic, not a realistic, motivation."

Carter: "Wilson is always defined as an idealist in U.S. foreign policy."

Elizabeth: "But note it would mean that descriptive realism is false: it would be false to say that states, or governments, or countries, are *never* motivated by moral values, but rather, only by values of rational self-interest. Arguably, these two examples show cases of countries acting *against* their self-interest, but because of their conception of morality."

Lori: "You know, I've even thought that that is what it means, truly, to have a moral commitment, to attach a moral value, to anything."

Nick: "What?"

Lori: "Precisely your willingness to give up something in terms of your interests for the sake of staying true to your moral value."

Elizabeth: "I see, like you give up the pleasure of having sex with somebody new and different for the sake of keeping your moral commitment of loyalty to your current partner."

Lori: "Bingo."

Carter: "Michael Walzer has a further important point here. He says any society which was run strictly on the basis of realism would not survive for long."

Nick: "Why not? Realism is all about selfishness, self-interest, and survival."

Carter: "Because, he says, a political community with only self-interest and survival as its purpose would fail to inspire the blend of loyalty and commitment necessary to sustain the political community, as it were. People need to feel part of something beyond mere self-interest and survival if they are to maintain a robust motivation towards keeping something going."

Nick: "Interesting, interesting. I suppose it depends on how complex you view human beings. If you see them as simple, then you're going to see them as satisfied with realism. But if you see them in a more complex light, then you won't."

Elizabeth: "But, clearly, there's so much more to human nature. While security and power are manifest concerns, we are also interested in so many more things—ideas, love, music, education, and so forth—above and beyond just basic security and increasing your income. Realism, at the very least, is a horribly reductive theory."

Nick: "Maybe, maybe. At least that's true of descriptive realism, distilling all the variety of human experiences and ambition into the claims about power and selfishness."

Carter: "Plus, I've always felt that interests and ideals don't have to be seen as competing opposites, right? Surely, it's in your interests to see your ideals realized in the world, no?"

Lori: "Obviously, American foreign policy sees it that way, right, or else why would the United States be trying so hard to create commitment to human rights, to democracy, and to economic growth . . ."

Elizabeth: ". . . and to bad TV . . ."

Lori: ". . . and to bad TV—hah—around the world? You're right, Carter; is the whole realism/idealism debate based on a false dichotomy—an untrue distinction—between interests and ideals?"

Carter: "Perhaps: it has always seemed to me that one's ideals should be consistent with one's interests. No point believing in something which is going to require you to kill yourself, right? And also that one's interests include one's ideals, such that one defines one's own happiness in terms of the degree to which one sees the world coming to fit one's values."

Lori: "Here's my big problem with realism: it's a self-fulfilling prophecy, right?"

Nick: "What do you mean? Prophecy? That's just Christian Realism, right?"

Lori: "No, not literally biblical prophecy. This is what I mean: realism recommends a certain way of acting and behaving, right? Purely selfishly, and with regard to one's own interests, right?"

They all nodded.

Lori: "Well, then, what kind of behavior is that going to draw forth from others?"

Nick: "Behavior which is just as selfish and self-serving."

Elizabeth: "Which will, in turn, make it irrational for you to behave in any way which is other than selfish and self-serving."

Lori: "And then what happens is that realism *creates* what realism already takes to be true—it creates this kind of ruthless, selfish, me-first world, where no one trusts each other, no one takes a chance in believing in idealism, and so on. So, it's not a description at all. It's a prescription which, if widely believed in, *would become* an accurate description of the world."

Nick: "But what's wrong with that, Lori? Are you urging national governments not to take care of their own interests?"

Lori: "I'm saying that it's set up against—biased against—idealism in the first place and, as a result, makes the world worse off."

Nick: "Whereas I say it contributes to making the world more safe and secure. We don't have stupid 'Don Quixote types' tilting at windmills, trying to create revolutions and empires and utopias, when no such things can exist in our flawed world."

Carter: "And you know that's actually one of the realists' most interesting arguments. Kissinger—a prescriptive realist—says we should all leave our moral and idealistic values at home when it comes to foreign policy issues like war."

Elizabeth: "Why?"

Carter: "Because people get hot and bothered, and emotional, about their moral ideals and, as a result, they make dumb decisions which aren't in their interests: like Don Quixote charging at windmills, only to come to defeat. Consider precisely something like war. David Welch, in his book *Justice and the Genesis of War* actually argues that what he calls 'the justice motive' produces more wars."

Elizabeth: "What?"

Carter: "Yes, he says that people like you, who believe there is a link between justice and war, actually make war more frequent, and more destructive, than the realists who are totally cynical about war. So, as you noted, people like President Wilson spurred the country on to involvement in a European war—WWI—not because it was in American interests but because of his own moral values. And was that a good idea? Was it in America's self-interest?"

Nick: "I once had one of my old professors almost spit with rage when he lectured us in class that Wilson was the worst U.S. president ever."

Elizabeth: "Why?"

Nick: "Because, he said—and I'm quoting—'without Wilson, there'd be no Treaty of Versailles; without the Treaty of Versailles, there'd be no Hitler; without Hitler, there'd be no Second World War.' I could hardly believe it—blame Woodrow Wilson for the Second World War."

Elizabeth: "That's, uh, extreme. The blame clearly belongs at Hitler's feet."

Carter: "But that's a common realist theme, whether descriptive or prescriptive; leave the moral values at home, locked up in the closet. When it comes to war especially, don't you dare even consider it in terms of what's 'just' or not; stick only to a sober calculus of practical rational interests."

Nick: "Seems wise. Everyone fights over whether their moral opinion is 'The Truth.' Shut up, sit down, and only discuss things like—is this in all our interests, yes or no?"

Lori: "No, think deeper, Nick. It's garbage. This view is suggesting that morality is itself a source of conflict and possible violence."

Nick: "Isn't it?"

Lori: "No. I mean, obviously, people can disagree about what's right and what's wrong. But those are different opinions about morality. The very function of morality, though, is to provide us with rules of conduct which can guide all our behavior, making all our lives better off. I mean, the law can't be everywhere, right? Where's a cop when you need one?"

Nick: "Amen to that."

Lori: "So we've created moral rules and values—don't lie, be courteous, and so forth—to organize our behavior and to minimize conflict when the law is not around. So, morality's job is *not at all* to create conflict: it is to *solve* conflict and coordinate human behavior in ways which benefit us all. We all benefit from a moral rule not to be cruel, or not to murder, and so on. So this realist view, that commitment to morality in war *produces war,* is ridiculous. It is a naïve, superficial, Balkanized view of ethics."

Nick: "Balkanized?"

Elizabeth: "After the Balkans region in southeastern Europe. Means fragmentation, and full of conflict."

Nick: "OK."

Elizabeth: "And Lori's point ties in with one I've been considering: isn't realism biased in favor of the powerful?"

Carter: "How so?"

Elizabeth: "Well, I think this is the reason why so many realists today are American. Since it's all about power, and since realism recommends you do what will augment your power, then what is going to happen in practice is a legitimation, or approval, of the dominance of powerful countries."

Nick: "Huh?"

Elizabeth: "Think about it. For realists, it's all about power."

Nick: "Yes."

Elizabeth: "And it's all about acting on the basis of interests, which are defined in terms of power."

Nick: "Yes."

Elizabeth: "Then, by that standard, who comes off smelling like a rose? Precisely the most powerful, because they can best do those things which express, and secure, their own power."

Nick: "So?"

Elizabeth: "So? What if they use their power in a bad way? That has been known to happen, after all! Realism degenerates into a pathetic form of power-worship, excusing all kinds of moral flaws and violations."

Nick: "Hum . . . and that would be especially problematic in wartime."

Lori: "You're damn right it would! I never thought of that. Realism would impose no moral restraints on how war is fought, right?"

Carter: "No, no *moral* restraints, because it denies that morality and war have anything to do with each other. But it would impose restraints of prudential self-interest. If you think back to our discussion of Iraq, I indicated how many realists thought it was too risky an idea, and advised against it. They would have the same attitude towards methods of war: only use, or do, what is in your self-interest. Avoid everything else."

Elizabeth: "But then that is a refusal to rule out actions which international law condemns as flatly wrong, like targeting civilians, or using rape as a tool of war, or using weapons of mass destruction even."

Carter: "That's true. They don't rule such things out as a matter of principle. But they would consider this flexibility a virtue: don't rule anything out; put everything into the mix and see what works best, strategically, in the situation."

Elizabeth: "But I think that's morally wrong. Some things should be totally ruled out, and not even considered as options in wartime, like the things I mentioned: WMDs, rape, torture, terror, civilian massacre."

Nick: "Whereas I think the flexibility of the theory is a virtue, and much more realistic on the battlefield."

Elizabeth: "But that kind of 'flexibility' is exactly what leads generals and soldiers to consider doing, and then to do, such horrible things. Far better to have firm and clear moral and legal principles that such things are off the table."

Lori: "I agree with Elizabeth."

Carter: "Whereas I agree with Nick. I should point out that it's hard to see many circumstances where doing those terrible things will be in anyone's enlightened self-interest. In practice, I think there's much overlap."

Elizabeth: "Yet it remains a real difference in principle: should these things even be thought of? International law says no, whereas realism says yes."

Lori: "And you know Carter, your assurances about this not often happening in real life, they don't ring true to me. It seems you have this idea about a very smart, conservative, cautious realism. Yet you admit that, in the past, realism's 'balance-of-power' policy produced more wars than it prevented. Some realists have a very aggressive definition of risk and it leads them into wars, and to do things in wars, which backfire and create bad consequences. To prevent that kind of risk and variability, we should just do what Elizabeth says; have firm and absolute rules which just forbid such activity. It's so much easier and clearer and trust-creating—or stable, anyways—than a system which always says 'well, it depends . . . .'"

Nick: "Not bad, not a bad critique. For me, what does it is the bias in favor of the powerful. That's clearly there in realism."

Carter: "I must admit this has me thinking, too. When you think in terms of being inside a tradition for so long, it rattles you to be forced to think in terms of challenges to its most basic assumptions."

Lori: "Don't get me wrong. If you are fighting to survive, I think realism is right on, totally unavoidable. Your ultimate self-interest is not to be conquered or annihilated, right? But all these other wars, like Iraq, this is where I see realism causing lots of problems."

Elizabeth: "Absolutely. I think this whole historical reference to the balance of power doctrine is very apt, and continues to apply to contemporary realism: there's this huge scope for disagreement about, and serious mistakes regarding, what is in your best interests to do at any given moment. I mean, as a principle, it's smart and easy and who could deny that? But during practical application, especially during war, there's a big risk of evaluating the rules wrongly, and your interests wrongly, and you end up making enormous mistakes you can't make up for. Far better to be guided by unchanging and clear general rules—like in international law—than to have to do this constant, shifting calculus of trying to guess what's in your best interests."

Nick: "Would you be willing, Elizabeth, to tell us more about international law's attitude towards war? Like, expanding on the meaning of these clear, general, and absolute rules? I mean, as a General, I know quite a bit about them—especially rules of engagement and battlefield rules—but I'd appreciate a top-to-bottom review and refresher."

Elizabeth: "I'd be very happy to do that."

Carter: "For my part, I'm not totally willing to give up on the wisdom of realism. Maybe it does, or can, backfire. Maybe it is biased towards powerful countries. Maybe it should rule out weapons and tactics that it doesn't. But the core principles seem so true to me that I can't just give it up. Most countries do behave fearfully and selfishly, and it would be foolish to ignore that. And power is an irrefutably useful good to have, no? I'm most impressed with a cautious, prescriptive form of realism which doesn't deny morality, and which views rational risk in a sober, conservative way and not a reckless, aggressive way. I think such an approach is ethically principled, and is the least likely to backfire, and is the most likely to lead to smart decisions about war: we should only pull the trigger when our most vital interests in survival and basic needs obviously demand it. And, once we're in, we fight to win—of course—as it's never in the national interest to lose a war. But we win well. We choose our weapons, tactics, and battles in such a way that we remain civilized and restrained and we don't antagonize world opinion. We don't fight in a way doomed to wreck the peace. But the key wisdom of picking your battles wisely, tending to your rational self-interest, and not being too naïve and trusting, these will always make me a kind of realist."

Nick: "Well, that's very well said on realism's behalf. Or, at least, that kind of realism. Me, I guess I'm looking for a kind of blend between realism and idealism: smart yet principled; self-protecting yet also concerned sincerely with trying to make the world a better place."

Carter: "I once heard Richard Falk, a political scientist, talk about 'visionary realism.'"

Nick: "Interesting."

Carter: "This would be a realism based on pretty cynical views about current human and government behavior yet unwilling to believe we're trapped in a cycle of power-seeking, violence, and war. There's a commitment to improving the world. There's a vision about how the world would be better off. But that improvement must come—can only come—through realist means and methods."

Lori: "Which would probably involve imposing such a vision through the exercise of power."

Carter: "Yes, but they see that as the only feasible way in our world."

Lori: "But then we're back to the pro-American empire people, no? They have a vision—of democracy, capitalism, human rights, U.S. hegemony—and they're seeking to coerce the rest of the planet to act in line. Visionary realists? Or old-fashioned imperialists?"

Carter: "I'd prefer the former, but am aware of the risks of, and similarities to, the latter."

And at that, the tour of the base was over and they all returned to the hospital.

Chapter Five

Marching Off to War

Carter Johnson left to do his duties and, since there were as yet no medical updates, Nick, Lori, and Elizabeth sat in Gil's room—late afternoon, coffees in hand—and renewed their discussion.

Nick: "So, Elizabeth, you promised us international law's perspective on these issues."

Lori: "OK, but wait a minute. What is international law, exactly? I've always wondered that."

Nick: "You know, me too. I've met several so-called 'international lawyers' in my day, and I've always wondered what, exactly, it is they do."

Elizabeth: "Fair enough. Lord knows, I've met a ton of 'international lawyers' in my day, too. Sometimes I wonder what we're all doing, too, or why we bother."

Everyone laughed.

Elizabeth: "International law is, essentially, the sum total of a series of contractual agreements between nation-states."

Lori: "Wait, wait. Define your terms, please."

Elizabeth: "OK. I guess this is quite important, isn't it? International law is built up around the idea of a nation-state. Most scholars date its original founding to the signing of the Treaty of Westphalia in 1648."

Nick: "I've heard people speak of the 'Westphalian System.'"

Elizabeth: "Yes, sure. The 'Wesphalian System' refers, very generally and in an all-encompassing way, to the current system of international law and international relations. It is a system built around the omnipresence of the nation-state."

71

Lori: "'Omnipresence'?"

Elizabeth: "Sorry. The, uh, 'ever-presence.' The ever-present presence, so to speak, of the nation-state."

Nick: "And the nation-state?"

Lori: "Well, that we've already had, remember? The state is the national government, whereas the nation is a group of people who think of themselves as a people—a separate and unified group. United by a way of life and a shared historical experience."

Elizabeth: "Right. What's vital to note is that this way of organizing political life wasn't always the norm. There used to be imperial ordering—one city, or country, ruling a far-flung empire composed of many nations, or a kind of local or municipal ordering of a government ruling over just one part of one people in a tightly specified area. The dominance and growth of the nation-state is a kind of middle-ground between far-flung, cosmopolitan, imperial governance, on the one hand, and a very localized city-to-city, or region-to-region, governance on the other."

Lori: "What about this 'Treaty of Westphalia'?"

Elizabeth: "Signed by various kings and princes, on behalf of their countries, in 1648. Brought about an end to the Thirty Years' War, in 1648. This was a war of religion, of Protestant versus Catholic, in Europe. Very bitter and violent. The Treaty is credited with creating the modern system of international law—a law based around nation-states with rights to political sovereignty and territorial integrity."

Lori: "Can you define those, please?"

Elizabeth: "Sure. The Treaty of Westphalia was focused on securing the peace in Europe after a bitter war of religion, over whose interpretation of Christianity ought to prevail. The peace was secured, eventually, through the following proclamation: the nations of Europe agreed not to interfere with each other's choice regarding an official religion amongst their own people. Each state was recognized by every other as entitled to have sovereignty—or control—over the religious beliefs of their own people. Each state said, in effect: 'Hey, we won't interfere, and certainly not through force of arms, regarding what your people decide is the religion they prefer. We won't try to force your people to adopt our religion through force of arms. Your people can decide that for themselves and our people, in turn, can decide that for themselves, and we look to you to provide the same kind of respect and noninterference that we are offering to you.'"

Nick: "A good deal. Very sensible."

Elizabeth: "Arrived at only after thirty years of utterly senseless bloodshed, and the realization that neither side had the force, or the numbers, to convert the other. But you're certainly right, Nick, a good deal; mutual peace secured through mutual respect of each other's sovereignty."

Lori: "So, sovereignty means control?"

Elizabeth: "Nation-states, or national governments—like the government of the United States—are the most basic actors acknowledged in international law. In many ways, they are the only actors acknowledged in international law. Indeed, international law is essentially a series of contracts between nation-states. The nation-state arose as a kind of middle-ground solution, in the West, between the cosmopolitan, imperial dominance of the city of Rome, on the one hand, and the local tribal anarchy which followed the collapse of the Roman Empire, on the other. So, you see, from about 1000 AD onwards, the growth of these national groupings—big, extended 'tribes' based on shared languages, customs, ethnicity, religion, worldview, behavior, and so on. And these nations wanted to be governed by their own self-selected government, according to their own customs and values. This is what is meant by 'political sovereignty': the control, by a nation-state, over its own internal affairs. The right of its people to govern themselves by a manner of their own choosing."

Nick: "Makes sense. Freedom, right? Freedom from imperial control."

Elizabeth: "Exactly. The freedom of a defined group of people. But that freedom must be exercised *somewhere*, right? And so the right to territorial integrity co-exists with political sovereignty as the two most basic rights of states acknowledged by international law."

Lori: "I've always wondered about the difference between these two rights."

Elizabeth: "Well, the logically, legally, and morally prior right is that of sovereignty. The right of a people to rule itself free from imperial control The right of nations to self-determination, or political self-rule, so to speak. A people ought not to be forced into a system of governance to which it does not agree."

Nick: "Absolutely. The American Revolution was motivated by such a desire, such a principle of freedom."

Elizabeth: "True. It's the basic moral building block of modern international law. But, to return to Lori's query, this basic building block—this foremost role of sovereignty—must be exercised or enjoyed by a group of people *somewhere*, no? And this has to be on land. We can't live in the air, or under the sea, or in the trees, and so we need land to live on and live off. People need land to exercise political sovereignty. And so this is where we get territorial integrity."

Lori: "How so?"

Elizabeth: "Because it is seen as a necessary condition for the exercise or enjoyment of political sovereignty in the real world. For sovereignty to mean anything, it must be practiced and respected within a concrete location—a specific chunk of land—as both a space to live and as a source of natural resources for the people to extract and use. You know, things like fertile soil to grow staple crops in, trees to provide wood, access to rivers and lakes, or well-water, to drink, and so on."

Nick: "So, the value of a people to enjoy self-determination within a clearly marked, and claimed, territory is the basic concept and moral principle at the bedrock of modern international law."

Elizabeth: "Yes. Well said."

Lori: "And international law is a series of agreements between states, on behalf of their peoples."

Elizabeth: "Yes. States negotiate deals between themselves, on the basis of mutual advantage and common interest. The deal is called a treaty. State governments then take that treaty back home to their own country, and pass it into law, the way they would any other bill. Once it passes, it is as much binding law as anything else in that society, like the speed limit, and so on."

Lori: "So that's how it's done."

Elizabeth: "Yup. No big deal, really. And international lawyers help draft those treaties to begin with, and they help interpret them after they've been implemented and enforced. Sometimes disagreements arise between countries over the meaning of certain terms or principles, and international lawyers have to argue it out."

Nick: "In court? Is there a world court? I thought there was no world government."

Elizabeth: "Well, there are some international court systems, but nothing like the One World Court option of which you're thinking. I'll spare you the details. The relevant thing, for our purpose, is that there are international laws and treaties regulating war. International lawyers refer to them as 'the laws of armed conflict.'"

Lori: "What? Ha! The laws of war?! You're kidding, right? I thought 'all's fair in love and war.' Once you're in war, you fight to win, and the winner takes all once it's over."

Nick: "Not true, Lori. I mean, many countries may still behave like that in practice. But, in terms of international laws, they are not supposed to do that . . ."

Elizabeth: ". . . and, don't forget, international law comes out of a promise of good behavior they themselves have made."

Nick: "Right, and so when they violate these laws, it's like breaking their own word, betraying their own commitment."

Lori: "Big deal. People break their word all the time."

Nick: "That hardly makes it right, Lori."

Lori: "But it does raise issues about the actual effectiveness of international law, doesn't it?"

Elizabeth: "It can, sure. Realists, for example, would probably say international law is only relevant and effective—that is, it will only be followed—if it is in the interests of the countries involved. If not, then it won't be followed, and we should expect betrayal and hypocrisy."

Nick: "But surely international law most often *is* in the interests of countries. After all, as you said, Elizabeth, international law is a freely-created, freely-chosen deal between countries. Counties wouldn't have agreed to the deal if they didn't expect it to serve their interests and benefit them in a real way."

Lori: "Right. I mean, I don't deny that. But countries, just like persons, can make mistakes about what is truly in their best interests. They often fail to behave in a way consistent with their best interests."

Elizabeth: "Well, I can hardly deny that. I'm just trying to lay out how international law views war. The rules and regulations it sets out in connection with armed conflict. I think these rules *are* in everyone's best interests and that's why they have found their way into international law. To that extent, it's an ideal and, as you note, Lori, both people and countries sometimes fail to behave in an ideal way, and they make mistakes about their interests."

Nick: "Sure. They get angry, emotional. Or they act on bad information or intelligence. They bet the future is going to turn out one way when it turns out another, and so on."

Elizabeth: "But that's precisely when international law is needed: to hold countries accountable to the higher standards they've already agreed to, back when they were thinking fairly and clearly."

Lori: "Fair enough. So, what are the international laws of war? What are—this sounds silly to say—the ideal ways for countries to behave during wartime?"

Elizabeth: "Well, there's a fundamental distinction in the law between going to war and fighting a war—that is between starting a war in the first place and then fighting it properly, once it has begun."

Nick: "*Jus ad bellum* versus *jus in bello.*"

Lori: "What? Speak English."

Elizabeth: "You know lawyers and their love of Latin, right? '*Jus ad bellum*' is Latin for 'the justice *of* war.' This refers to the rules governing when it is legally permissible for a country to start a war, or to resort to war by way of solving a problem."

Lori: "What kind of problem?"

Elizabeth: "Essentially, international aggression. Let me come to that. The other legal term Nick used was '*jus in bello*.' This is Latin for 'justice *in* war.' It refers to the many rules and regulations governing how countries may fight once war breaks out. The first set of rules is the responsibility of those who start wars: usually, the head of state or whoever holds the 'war power' in that country, for instance as laid out in that country's constitution. They must ensure that their nation fulfils *jus ad bellum*, or else they may face war crimes charges after the war."

Lori: "Has that ever actually happened?"

Elizabeth: "Yes, but only rarely. Nuremberg in Germany in 1945. And then again, very recently, in connection with the civil wars in the mid-1990s in Yugoslavia and Rwanda. The sad truth is that heads of state almost never sit on trial for setting unjust and illegal wars into motion."

Nick: "Whereas officers and soldiers get put on trial all the time for *jus in bello* violations."

Elizabeth: "You're right, there's quite an imbalance. Whereas *jus ad bellum* is the preserve of those with the 'war power,' *jus in bello* is the responsibility of those who fight the war set in motion by the political leaders; the soldiers and the officers who command them."

Nick: "Like me."

Elizabeth: "Correct, General."

Nick: "OK, well, I know all about *jus in bello*—I'd better, right?—so I'd rather we focus on *jus ad bellum*, if that's alright."

Lori: "Whereas I wish to hear about both."

Nick: "Ok, but let's start with *jus ad bellum*."

Elizabeth: "Alright. The most relevant piece of international law here is the Charter of the United Nations, first signed in 1945 and now signed by very nearly every single country on Earth, about 200 of them. Note that, when you sign on to this treaty or become a member country of the UN, you are agreeing to its principles and promising to behave in a way consistent with the Charter. You are freely accepting these rules and regulations."

Lori: "Right."

Elizabeth: "*Jus ad bellum* stipulates that, as the one with the 'war power,' one has a responsibility to ensure that when resorting to war, one stays in line with the following rules; just cause, proportionality, last resort, and public declaration of war by a proper authority."

David Pearson: "Don't forget the rules of right intention and probability of success!"

Elizabeth: "Hello, I'm sorry I don't believe we've met." Elizabeth rose out of her chair and shook David Pearson's hand. He was mid-30s, still quite fit, with dark hair and dark eyes.

David: "I'm David Pearson. I'm here to see Gil. I used to be his professor at West Point. I'm also a just war theorist, and that's my explanation for my rude interruption. Please forgive."

Nick stood up and shook David's hand.

"Nice to meet you, Professor Pearson. I'm General Nick Gordon, Gil's uncle. Have you come all the way from West Point just to see Gil?"

David: "Please call me David, General. No, I actually haven't. I was giving a talk at a conference at the University in Jena in the middle of Germany. I was informed of Gil's injuries via e-mail by another former student of mine—one who was in Gil's class, actually—and then made my way down. I always got along very well with Gil, and wanted to check on his well-being. The e-mail suggested the injuries were serious, but didn't really go beyond that. Gil was caught in a firefight?"

Elizabeth: "And a roadside bomb explosion. In Baghdad. I'm Elizabeth McAllister, JAG core, based at the Pentagon. Pleased to meet you."

Elizabeth proceeded to explain her professional reason for being on the scene.

Elizabeth: "And this is Lori Gordon, Gil's wife."

David: "Pleasure to meet you, Lori. I wish it were under different circumstances. Gil was a very friendly, pleasant, and bright student. Are his injuries very bad?"

Lori: "Yes, apparently. Focused on his head and neck. Life threatening, even. The doctors are doing various things to him right now as we speak."

David: "I'm so very sorry, Lori. And you are both so young. I don't know what to say beyond that."

Lori: "That's alright, Professor Pearson. I really appreciate your coming at all, and I know Gil would, too. Very thoughtful. But, look, Gil is with the doctors and surgeons, and I've no idea when they'll be done doing whatever it is they're doing with him. They haven't updated us for hours. So, we've been killing time, talking about this stupid war in Iraq, and the conversation has moved on to war more generally. Won't you join us? It seems like you were eager to do so anyway . . ."

David: "It would be my great honor, thanks. So, from what I caught there, you were discussing *jus ad bellum*?"

Nick: "Yes, Professor. Now you seemed to want to add to Elizabeth's list of rules. Why?"

David: "Well, from what I heard, General, she was explaining the current set of international laws regarding *jus ad bellum*."

Elizabeth: "That's correct."

David: "And I just wanted to add the other two rules for *jus ad bellum*, which just war theory insists upon, in addition to those which international law endorses."

Nick: "There's a difference between just war theory and international law? I thought the law grew out of the just war tradition."

David: "International law—I gather you mean post-Westphalian international law—draws very heavily on the just war tradition."

Lori: "I've never heard of 'just war theory,' or the 'just war tradition.' Can someone set me straight, please?"

Elizabeth: "Please go ahead, Professor Pearson."

David: "Well, I will, but only very quickly, and then I'll turn it back over to you. I'm sorry I interrupted you when I walked in. Just trying to break the ice."

Elizabeth: "Not at all. Please tell us about the relationship between just war theory and the international laws of armed conflict."

David: "Well, as I understand it, these international laws have grown out of just war theory."

Nick: "That's how I understand it, too."

David: "The just war tradition, or theory, has an ancient pedigree, dating back at least to ancient Greece. It's a shared body of thought about the ethics of war and peace which has developed over centuries. The core principle is that it is sometimes legitimate, or morally correct to resort to war. In particular, it is OK to do so as an act of self-defense against aggression. Just war thinkers developed the rationale and justification behind this mode of thinking and defended it against pacifists and realists of various stripes."

Lori: "We just had a long conversation, outside while walking around, about realism's attitude toward war."

David: "I'm sorry I missed it. Realism is deeply mistaken about what to do about war."

Nick: "So, just war theory knows better?"

David: "Yes, I dare say. Realism's pessimism is just too dark, and it ends up calling forth precisely the kind of cynical, relentless violence it swears it deplores."

Lori: "Yeah, we covered that. Forgive my rudeness, but let's not go over ground we've covered."

David: "I'm sorry."

Lori: "Not at all. You weren't there. Perhaps you could help us by clarifying the relationship between just war theory and international law, and then we can return to Elizabeth's explanation of the international laws of armed conflict."

David: "Sure. Just war theory stretches back thousands of years to ancient Greece, whereas modern international law didn't begin until 1648. When international lawyers were crafting rules for the regulation of armed conflict, they didn't want to reinvent the wheel. They were thinking: are there common, well-understood principles of just conduct when it comes to wartime behavior? The answer was yes, and it was those principles articulated within just war theory. So, when drafting treaties for nation-states to agree to, international lawyers—like Hugo Grotius—drew heavily on the preexisting moral, philosophical, and even religious principles which had gelled over the centuries into just war theory."

Nick: "So, it would be fair to say that international law and just war theory have a lot in common—that international law draws upon just war theory—but that there are also some differences?"

David: "Exactly. Like in *jus ad bellum*, for example. International law does not include the rules of right intention or probability of success."

Lori: "OK, but stop there, please, for now, as we haven't heard from Elizabeth regarding the content of the other *jus ad bellum* rules in the first place. What were they, again? Just cause . . ."

Elizabeth: "Just cause, proportionality, last resort, and public declaration of war by a proper authority."

Lori: "And a just cause for war is resisting international aggression, you said."

Elizabeth: "Yes. Just cause is really the key to everything."

David: "Absolutely. The most important rule, by far. It sets the tone for everything else which follows."

Elizabeth: "And the key concept, from international law's point of view, is the resistance of aggression. The commission of aggression is really what gets everything in motion. It is the crime, so to speak, which sets the stage for war."

Lori: "As a punishment for the crime of aggression?"

Elizabeth: "Sort of, though many people don't like that wording. They prefer the terminology of 'resisting aggression.' Aggression gets committed, and war is authorized to resist and punish aggression, to defeat the aggressors and to nullify—or turn back—their gains."

Lori: "But what's aggression?"

David: "That's the million-dollar question."

Elizabeth: "But one thing I like about international law, say in contrast to just war theory, is that it has quite a clear and focused definition of aggression. The Charter of the United Nations, Article 51, is the key here for the authorization, or permission, to go to war. Aggression, it says, is the use of force against another member of the UN."

Nick: "But what's 'force'?"

Elizabeth: "Physical force. More specifically, the deployment of one country's military assets—army, navy, air force, marines, coast guard,. weaponry (like missiles)—across a recognized border and into the territory of another country."

Lori: "That is pretty specific."

Elizabeth: "Clear, huh? So, it's not aggression if you're merely yelling at someone, or angry at another country, or even if you're taking more serious moves like imposing sanctions on another country. It's only aggression if physical *violence* is involved, like sending soldiers across a border to invade some other country's territory."

Lori: "But let me play devil's advocate here, for a moment: why is that wrong?"

Elizabeth: "Precisely because it violates those two foundational principles contained within international law: the political sovereignty and territorial integrity of a nation-state. A people has the right, within its territory, to rule itself according to its own principles. Sending an army into that people's territory is done precisely to interfere with its freely chosen ability to determine its own fate, right? You send the army into the other country to take it over, or at least to coerce that people into giving you whatever you want, correct?"

Lori: "I guess that's true."

Elizabeth: "So, aggression violates the most basic values of international law. That's why it can be resisted and punished with something as forceful and serious as warfare."

Lori: "OK, let me play dumb for a moment. Why is national self-determination so amazingly important that its violation can justify *war* in response? I mean, we're talking war here, right? War, the single most destructive practice which humanity has yet devised. We're willing to authorize and unleash this just because we don't want some strong country taking over a weak one? I mean, c'mon, is that truly worth it? Is it worth the cost?"

David: "Absolutely it is, Lori. Here is where I think just war theory has something further to say, something substantial to add. International law only specifies or stipulates *that* something is wrong or prohibited, whereas just war theory can more fully explain *why*."

Nick: "OK, Professor, please do so."

David: "Think of a world wherein aggression is allowed to happen, and is neither checked nor resisted nor punished."

Nick: "It would be horrible. Everyone would be committing aggression against everyone else. I mean, why not, right? They might get away with it, and get to keep the gains of their aggression. It would be, as Hobbes said, 'the war of everyman against everyman.'"

David: "Precisely: the law of the jungle. Exactly the law of the jungle. There would be no rules or order, as everyone would be free to commit aggression against everyone to their heart's content. This is to say that the horrible wrong which aggression commits is the violation of the very possibilities of human civilization itself. We need to be able to count on not suffering aggression if we are able to do anything meaningful with our lives at all. We need that level of security, and respect for our rights, to rule ourselves, if we are to progress from the terrible ways of the jungle and to enjoy a rational and orderly progress of life which fits our being human."

Lori: "Wait, you're saying that war is a way we stick up for our rights? War is a way we *escape* from the laws of the jungle? Looking at it, it seems to me that war is a way we revert from peace *back into* the laws of the jungle."

David: "No. It depends on who started it."

Lori: "That's something a child would say."

David: "Because a child grasps the basic moral truth of it, and doesn't lose that truth amidst all the conceptual clutter we adults put up around it."

Lori: "Say more."

David: "Well, we're all entitled to our lives, and to our life-plans, right? Whether we are persons or nations. This is a hallmark of our humanity. When someone uses physical violence to threaten our lives, or our freedom to choose what we do with our lives, then they strike at our most basic human rights and our most elemental humanity. *They seek to dominate us as if we were their slaves*, and to do so through force as opposed to changing our minds through persuasion. Aggression is the utter denial and negation of humanity. It seeks to drag us all back to the jungle days of our original ancestors. As such, it seeks to drag us back, cancelling all our progress, all our rules of civilization and, of course, in violation of our most fundamental rights. The person or country who initiates this kind of process is committing a terrible injustice."

Lori: "But why isn't the person or country who responds in kind committing a similar injustice? Two wrongs don't make a right."

David: "Because their response is merely what's needed to respond effectively, with equal or better force, to the initial violation. The initial violator is the 'Aggressor' whereas the one who responds is the 'Defender.' *It is never wrong to use force in defense,* providing force seems the only plausible way to resist or prevent the aggression. And kids know it to be true: the one who starts it bears all the blame, and must accept the consequences. All those who respond to the initial Aggressor, they merely seek to punish him and to restore the initial order disturbed through his violation."

Nick: "You know, I've never really thought of it in those terms. But you're completely right—aggression violates our most basic human rights, and would drag us all back to the law of the jungle. I used to think of it merely as a violation of freedom, but now I see it's so much more serious than that."

David: "Well, freedom is very much part of what we mean by the essence of humanity, and the escape from criminality. Freedom from aggression. Freedom from rights-violating force. Freedom from brutal violence which seeks to dominate. Freedom to determine one's own life and future. I think we are actually talking about the same things, in the end. It's the value of these things which international law and just war theory seek to protect."

Lori: "So, the only just cause for resorting to war is defense from aggression?"

Elizabeth: "Article 51 of the Charter says that, when aggression happens, two kinds of defensive wars are justified: 1) a war of self-defense, on the part of the victim of aggression; and/or 2) a war of other-defense, on the part of any third country, or set of countries, rushing to the aid of the victim."

Lori: "So, any nondefensive war is unjustified? Offensive wars are illegal?"

Elizabeth: "Not necessarily. Don't get me wrong, both the law and just war theory are crystal clear: wars of defense against aggression are the most justified and the least controversial."

Nick: "Which is just plain common sense; everyone knows you may defend yourself from attack, or come to the aid of someone else being attacked."

David: "True of both persons and nations. Aristotle, I think, was the first to speak of wars of defense being 'morally obvious.'"

Lori: "And nondefensive wars?

Elizabeth: "International law says that, for any war falling outside the definition of defense, prior authorization of the war action by the UN Security Council is required."

David: "Whereas just war theorists are split on the issue."

Lori: "Ah ha! Like this damn Iraq War, right?"

David and Elizabeth nodded.

Lori: "So, lemme guess: The Afghan war was seen as a defensive war, since America was attacked on 9/11, and so America and the Allies going to war against the Taliban regime was seen as a justified defensive response to, or punishment for, 9/11."

Nods all around.

Lori: "But then Iraq *didn't* attack us, did not commit aggression against us. And so—ah ha!—that's why they had that big showdown before the war at the UN Security Council: America was trying to get Security Council permission, and they never got it. So, it was an illegal war! That bastard Bush!"

Elizabeth: "Well, not so fast. Technically—legally—the United States withdrew its quest for Security Council authorization before a vote was held. But, obviously, it did so because it knew it was going to lose the vote. The United States then fell back on two further arguments to justify that war: one legal, the other political."

Nick: "What was the legal one?"

Elizabeth: "The legal one was this: the authorization to go to war in 2003 was contained in a prior UN Security Council resolution—the one in 1991, which described the terms of peace for the first Iraq war. That peace treaty clearly said: 'Iraq will abide by the terms of peace and, if it doesn't, then America and Britain reserve the right to resume hostilities as a way to bring about enforcement.'"

Nick: "But was Iraq in violation of the 1991 treaty?"

Elizabeth: "Repeatedly throughout the 1990s. Iraq kicked out UN inspectors who were searching for weapons of mass destruction (whereas the treaty required full Iraqi cooperation); Iraq flew military craft into the 'no-fly-zones'; Iraq failed to use food aid money on actual food; and so on. It is generally understood, historically, that, when you violate a peace treaty you freely signed, you open yourself up to armed resistance in search of forcing you to comply with your word."

Nick: "News to me."

Lori: "Me, too. That's actually a pretty good argument. Hmm . . . but we all know that *wasn't* the argument the Bush Administration used to justify the war. It was this argument of preemptive strike, or 'anticipatory self-defense.' The whole mess with the weapons of mass destruction thing, which turned out to be bogus. We've already spoken of this. Preemptive self-defense, my ass."

David: "Does it have to be bogus, though?"

Lori: "Well, explain to us how the just war tradition has been split on this issue."

David: "Well, there are some, let's call them 'strict defense purists,' who insist that *you must wait to be struck first* before you can strike back. Defense is defense is defense; plain and simple. Then, once you, or your ally, have been attacked, you may resist and punish with force. No muss, no fuss, no room left for difficult cases; it all becomes quite clear. If you attacked first, you're the aggressor. You're unjustified and illegal. Everyone else is justified in fighting you off and punishing you. War is just too serious and severe a thing to permit under any looser conditions. Vitoria—one of my favorite thinkers in this regard—said that, to wage war against a country which has yet to attack is like 'punishing a man for a sin he has yet to commit.' I love that line. You've got to wait for the attack before you resort to force."

Lori: "And so Bush violated that principle."

David: "Indeed. But that's only the understanding of the strict defense purists. Others believe that, if you look more deeply at the principle of defense, you'll discover the permissibility of some kind of first-strike principle."

Lori: "Really? And who would these people be? The people in Bush's Administration? Cheney, no doubt?"

David: "Well, perhaps. But also others with no connection to that war. Some medieval just war theorists, for instance, or Michael Walzer."

Nick: "Walzer! How many times have I had to read him?"

David: "You're not alone. Walzer's *Just and Unjust Wars* is considered the classic contemporary statement of just war theory, and it's in its fourth edition now. Impressive."

Nick: "Walzer supports anticipatory attack?"

Lori: "I guess you've read him often, just not remembered him, right, Nick?"

They all laughed.

David: "Yes he does, under exceptional circumstances. The key concept for those just war theorists who do support preemptive strike is this: *protection.* Clearly, defense is about protecting, right? Protecting one's people from the harm caused by an aggressor. Well, suppose you're the head of state, and suppose you've got overwhelming evidence of an upcoming invasion, or terrorist attack. As the head of state, you're duty-bound to protect your people from harm. So, what responsible president, or prime minister, is going to sit on his or her hands and wait for the upcoming attack, when, if he or she strikes first, that might prevent the plotted attack from ever hitting its intended target? It can be consistent with defense and protection to launch a preemptive strike."

Nick: "The best defense is a good offence."

Lori: "But the key move there was the claim about the 'overwhelming evidence of the upcoming attack.' How certain does that knowledge have to be? I mean, we can probably imagine many presidents, paranoid ones, saying: 'Oh, well, I'm sure that traditional enemy of ours over there is plotting something. So, better safe than sorry: drop the bombs on their head!'"

David: "And that is, admittedly, a real difficulty. Leaders might be hypersensitive, they might perceive threats where none are real, or the evidence they are presented with might be biased or presented towards them in a biased way."

Lori: "Like Bush. As president, he must have been simply terrified of another 9/11 happening on his watch. And so he wasn't willing to take the risk that Iraq wouldn't attack, or didn't have weapons of mass destruction, or didn't plan on giving them to al-Qaeda for use on America, and so he perceived the threat in an exaggerated way, and ordered the war."

Nick: "I thought you were of the view that Iraq was, and is, all about oil, Lori."

Lori: "Well, I'm just trying to put myself in the president's shoes. Strictly from the point of risk-management, I can see how his perception might lead to the logic of 'strike first.'"

David: "But I've got two major problems with the president's decision to attack Iraq in 2003."

Nick: "What are they?"

David: "First is the fact that he just surrounded himself with a bunch of 'yes men.' With the possible exception of Colin Powell, all the people around the table were of the same view about the war and then a kind of 'group-think' process took over, wherein they all just confirmed each other's fears, exaggerations, and misconceptions. The president should've selected advisors with different views on Iraq, the Middle East, and the evidence, and the result would've been a more balanced, comprehensive look at the data he needed to make a better decision. And, since I think this sort of thing—letting your emotions (especially fear) cloud your judgment—is very hard to avoid, I myself am a strict defense purist. No war, unless you've been attacked. It's just so much simpler and straightforward, with little room for abuse."

Nick: "Whereas I favor the 'best defense is a good offence' policy, and support the war decision. It was, as Lori said, actually a sound application of risk-management: couldn't risk another 9/11, so overthrow the regime who might be plotting another. Even if you're wrong about the weapons and the plot, you'll at least have gotten rid of a dictator. It all adds up to a green light for Go!"

David: "The second problem I have is that, as I understand from Bob Woodward's best-seller *Plan of Attack,* there was poor consideration of the postwar scenario. And this actually ties into the other *jus ad bellum* rules."

Lori: "How so?"

David: "Well, when I rudely burst in on Elizabeth, I mentioned probability of success. Just war theory requires you to show, *prior to the war's beginning,* that you'll, on balance, probably be successful in achieving your war aims. Whereas have we been successful in the postwar situation in Iraq? Indeed, what have our war aims been? This becomes so much more problematic when you do first strike. If you're a strict defense purist, and you've been attacked, the major war aim is simple: resist, repulse, and punish the attack. Whereas you get into all the problems of defining 'success' in the postwar situation when you strike first."

Nick: "Well, the goal was to remove the regime."

David: "Yes. But to replace it with what?"

Nick: "A democratic regime, elected by the people, that does not support terrorism."

David: "But how to achieve that goal? What are the lower-level goals, so to speak, needed to be achieved in order to secure that higher-level goal? That's what we didn't know, or at least planned so poorly for, at the outset of the war."

Elizabeth: "International law doesn't require that countries satisfy the rule of probability of success prior to the war."

Lori: "Why not? It seems very sensible: don't start a war which is going to be pointless; make sure that, on balance, you probably will be able to secure your just cause."

Elizabeth: "For two reasons. First, it's thought to be biased in favor of powerful countries, since smaller countries are going to have a harder time showing they can succeed, right? And why should smaller countries have a reduced entitlement to resist aggression just because they're small? It's not fair. The second reason is that the principle is forward-looking, requiring projection into the future, and that is going to involve speculation. It's just so much subjective guesswork involved in trying to predict how a whole war is going to pan out, whereas the law likes being as objective as possible."

Nick: "David, what was the other just war principle which international law leaves out?"

David: "Right intention."

Nick: "Oh, yeah. Augustine, right?

David: "Right."

Lori: "Wait, I don't know who Augustine is, or what right intention is about. Someone please explain."

David: "Augustine was one of the founders of just war theory. He lived around 400 AD and was a Catholic bishop when the Roman Empire started to collapse in the West. Two huge problems he wrestled with, personally, were these: as a bishop and Christian, he thought he was commanded to think and act, as well as to preach and teach, like a pacifist. I mean, look at Jesus' life and teachings; no real endorsement, or permission, allowed for violence, much less war. Love your neighbor—even love your enemy—and then turn the other cheek. Jesus himself didn't resist his arrest or crucifixion. So, Augustine thought Christianity mandated he be a pacifist, never permitting or engaging in war. But, as I mentioned, at the time Rome was falling apart, being attacked on all sides by a bunch of non-Christian barbaric tribes. It actually looked like the world was coming to an end. So, as a political thinker and figure, Augustine wanted to justify the use of force, in the name of Rome repelling these attackers, preserving Christian civilization, and preventing the unbelievers from taking over."

Lori: "So, how did he square the circle?"

David: "Through this rule of right intention. Augustine's conclusion was this: a Christian ruler must show love for, and hence protect, his own people. Thus, a ruler *cannot* be a pacifist; the duty of protection prevails. This is what justifies the use of force. But, to keep the ruler focused on achieving this sole permissible aim in war, a ruler must order war only with the right intention. No other motives, other than protecting your people, are allowed to enter your mind and spur your behavior: not hatred; not strategic scheming; not love of violence; not revenge . . ."

Lori: "Not oil!"

David: "Not oil, and so on. Your only motive, or intention, in starting a war which is allowed is the one motive, the right intention, of reluctantly doing what's required to protect your people."

Nick: "Seems right."

Elizabeth: "But international law doesn't require proof of right intention, either."

Nick: "Why not?"

Elizabeth: "Well, again it has to do with the law liking objectivity and evidence, and the belief is that there's just never going to be enough evidence to show clearly that those with the 'war power' are actually 'pure in motive' and have no mixed motives playing into their wartime decision-making."

Lori: "But right intention does seem to be important, no? When we talk about the 2003 Iraq War, for example, everyone has a strong opinion about Bush's real intentions and motives: oil? Revenge? Fear of another 9/11? Political strategy for changing the Middle East? Bush's own desire to seem strong and forceful? We could debate endlessly about the intentions and aims of war leaders."

Elizabeth: "But in a way, that's the whole point; we all debate it, but there's no way to solve that debate. There's no clear evidence. So the law says you've got to stick with what you know: was there an attack? Yes or no? What's on the leader's mind? Who knows, so leave it be."

Nick: "That makes sense, too."

Elizabeth: "Not that right intention isn't *morally* or *ethically* important. Presumably Bush knew what he had in mind, and he has to live with that. But the law can't really access that, and so the rule of right intention gets passed over."

Lori: "What are the remaining rules?"

Elizabeth: "Proportionality, last resort, and public declaration of war by a proper authority. Last resort, I guess, is the easiest; only resort to war when other reasonable means of solving the problem have failed. Diplomacy and sanctions are always trotted out as the morally better method of problem-solving prior to war: talk it out; try to cut a deal; bring in allies to gain influence, or leverage, on the enemy state; withdraw your ambassador; close your embassy; let them know how displeased you are. Then, if things really heat up, maybe cancel a treaty or two between your countries, kick out their ambassador, slap some sanctions preventing trade between your countries, maybe threaten war, and so on. War only as a last resort."

Nick: "Who could deny that?"

Elizabeth: "Indeed. Proportionality is somewhat related; only go to war if the problem truly seems so bad that war is a fitting solution. Proportionality, as in a proportion or balance between problem and solution. The rule of proportionality is supposed to force those with the 'war power' to think: is this problem *really so bad* that war is a good solution? Since war itself is so bad—and costly and risky—it's crystal clear that the only kinds of problem in view are very rare, very big, and very grave. Usually, there's a better way other than war."

David: "Sorry to interrupt, Elizabeth. But, to me, that is another consideration in favor of strict defense purism. With the preemptive strike people, how do you know the problem was really going to get so bad that force was genuinely required? You're jumping the gun, mainly out of fear. Whereas when you wait to see if you actually get attacked, well then it's crystal clear that war is a proportional response—as war has already been launched against you! Truly, I think that only defensive wars can ever satisfy all the just war and international law, rules."

Nick: "Interesting. Food for thought, for sure."

Lori: "But now: public declaration by a proper authority."

Nick: "The final *ad bellum* rule."

Elizabeth: "Well, there's two things here: the public declaration part, and the proper authority part. The first part is much easier; war must be publicly declared, by the proper authority for doing so, namely, that person, or branch of government, authorized with the 'war power' in that society."

Nick: "But, with public declaration, doesn't that detract from the element of surprise? From day one at War College, you are drilled with the notion that you always want the element of surprise on your side."

David: "That's Clausewitz, right? Speed and surprise as the keys to success in modern war."

Nick: "Also a bit of Sun Tzu: plan, prepare; plan, prepare. What's his big line in the *Art of War*? 'Every battle is won before it is ever fought.' Try to avoid war if you can, but if war you must, make sure you've got all the elements, and the factor of surprise, on your side."

Elizabeth: "But public declaration doesn't have to detract from the strategic use of surprise. Consider, for instance, what both Bushes did at the start of both Iraq Wars: they got authorization to go to war publicly declared, but then gave Saddam a deadline: anytime after 12 p.m., on Day X, we reserve the right to launch a war against you. Everyone knew this, it was nicely declared, and in both cases, the deadlines came and went without satisfaction, and the United States was still able to retain the factor of surprise, since no one knew, after the deadline, exactly when the attack would come. I think with Iraq 1, it was the next day and, with Iraq 2, it was only a couple of hours afterwards."

Lori: "This public declaration requirement seems awfully quaint and formal. Very medieval, like reading out a parchment, or getting the town crier to shout: 'Here comes war!'"

Elizabeth: "People often say this: what's the point of public declaration in our day and age? As I see it, it has less to do with the enemy and more to do with one's own people; those with the 'war power' must admit when they are waging a war, especially in a far-off land which makes getting full and accurate information back at home difficult."

Nick: "You're thinking, of course, of 'Nam."

Elizabeth: "Yes, and even Korea too. The various administrations were clearly engaged in warlike activities, and scale of operations, yet refused to admit it. It was dishonest, and it ties in very substantially to the proper authority aspect. In Vietnam in particular, there was a power struggle in the United States between Congress and the president—especially LBJ—regarding authorizing the war, and so one way the Johnson Administration—and Kennedy's too, I think—tried to get around that was to refuse to admit they were engaging in war. 'We're just sending military advisors, some weapons, just helping the local faction, and so forth.' But, if it looks like a duck, and quacks like a duck, as they say . . ."

Nick: ". . . it's a duck! Lordy, I remember those days all too well. But, Elizabeth, as a lawyer, say more about the war powers issue in the U.S. Constitution."

Elizabeth: "Well, if you look at it superficially, Congress has the power to authorize warfare: Congress, and Congress alone. Plus, Congress clearly has the responsibility, and power, to raise funds through taxes to pay for the military and to fund wars."

Lori: "But, if you look at it more deeply . . ."

Elizabeth: ". . . then you'll see that, in my opinion, the 'war power' is actually split between the executive and the legislature in American checks-and-balances fashion."

Nick: "How so?"

Elizabeth: "Because the president, as commander-in-chief, is plugged into the military chain of command. He barks, the military jumps. This gives him an enormous factual advantage over the slower moving Congress, particularly when it comes to responding to sudden, surprising events. Most recently, post-9/11, we've seen Congress extend sweeping war authorities to the president, precisely to empower the president in sudden moments of crisis, such as a terrorist attack. Or at least that's what happened when the Republicans controlled Congress. As the war became unpopular, along with Bush, and then the Democrats won Congress in 2006, they struggled once more over the 'war power.'"

Nick: "But they *should* have to struggle it out that way. The U.S. founding fathers were so smart; separate the powers and make them rivals, and then the only thing it becomes possible for government to do at all is the small number of things around which there is overwhelming consensus. Keeps government small and in check."

David: "But, recently, people have said that there should also be an *international* authorization requirement for war."

Elizabeth: "There is already. Well, sort of. To go to war for any reason other than defense against aggression, you must get prior UN Security Council authorization."

Lori: "Other cases like a preemptive strike . . ."

Elizabeth: "Or intervening in a civil war, or engaging in humanitarian intervention, or engaging in targeted strikes on suspected terrorist sites. Anything other than a military reaction to a first strike committed by some other country, either on you or someone else."

David: "Well, I've heard people argue that international authorization should be required even with defense. You know the phrase: 'one shouldn't be a judge in one's own case.'"

Elizabeth: "Ordinarily, I'd agree. But, when the attack has already happened, what's left to judge? Plus, the UN Charter goes out of its way to say that the right to go to war—to defend either oneself, or one's ally, against an armed attack which has already occurred—is an 'inherent right,' even 'a natural right,' of any state. No one should be able to take that away. For example, we might image an international body—like the UN—failing (for political reasons) to authorize a war of self-defense. Like if that country wasn't popular, or if the international body had its own internal differences of opinion. And, then, what if that country would just have to sit there and take it?"

Nick: "Outrageous. No way. We could easily imagine such an international body dealing with America in that way, or Israel."

Lori: "The real scandal, at least with the Security Council, was how it failed to authorize any intervention in Rwanda, and all those people got slaughtered. In 1994."

David: "People make a similar argument today in connection with Darfur, in Sudan."

Lori: "What exactly is going on there?"

David: "The country contains a Northern, mainly Arab and Muslim, population and a Southern, mainly black and Christian population. The government is controlled by the Arabs and it is permitting radical Arab groups to push the Christians out of the Darfur province. Like ethnic cleansing, of the kind seen in Yugoslavia, or even Rwanda; driving a whole people out of a region, or just flat out killing them. Very sad."

Lori: "And incredible to think it still goes on in our day and age."

Nick: "People always say that, but if they were to look at it historically, and honestly, they wouldn't be surprised, since it happens all too often and all too routinely."

Lori: "OK, so I guess none of us are too hot on the idea of an international authorization for war."

David: "Well, I myself think that, in theory, having yet another hurdle to jump over, before you get to go to war, is a good thing. But, I don't see how the Security Council really fills the bill; as Lori noted, it has failed to authorize interventions which should have occurred, but didn't."

Lori: "Plus, who the hell are these Security Council guys? What, or who, gives them the right to authorize?"

Elizabeth: "Well, remember, it's only authorization in the nondefense cases. When it's defense, each country simply has the right to go to war of its own accord. The thinking was that there may arise precisely the sort of preemptive strike cases, or humanitarian intervention cases, where force may be permissible. But, as David says, in such situations, we don't want people being judges in their own cases, and so we make them persuade the powers that be that their fears or concerns are objective and not subjective."

David: "I mean, there's sense in that."

Elizabeth: "But there's also these legitimacy issues dogging the Security Council. As Lori put it: who are these people? Well, the short answer is that they are the five major, winning powers at the end of the Second World War: America, Britain, China, France, and Russia. They set themselves, and international law, up as the final judges on war and peace issues. These are the so called 'Permanent Five' members of the Council, and they each have veto power over any proposed resolution. Now, there are ten temporary members on top of the permanent five, and these ten can vote in favor, but never veto, any resolution. The ten hold two or three year terms, I forget. In practice, though, the five hold all the cards and many wonder why that should be, especially when they are all northern countries (north of the equator) and developed countries (well, China is still partially developing). Three of the five are European, and four of the five are mainly white, or Caucasian, and have mainly Christian backgrounds. Should such an exclusive and powerful Council be peopled by countries with such a narrow representation of Earth's entire population?"

Nick: "Plus, there's just the practical issue of how many times it has failed in the past."

Lori: "Yes, failure plus exclusivity don't make for a recipe in favor of success, inclusion, and legitimacy. But another question I have with legitimacy is this: what happens when there is a dispute within one country, about who has the right to govern? Who has the 'war power' then? In other words, doesn't the legitimacy issue cut the other way, too?"

Elizabeth: "It does, but maybe more from the perspective of just war theory than international law."

David: "Yes."

Lori: "How so?"

Elizabeth: "In international law, you are the legitimate government if you are the one in charge in a given territory. It's a power test, really."

Lori: "But I thought that, if you're a member country of the UN, you're supposed to govern consistently with its ideals?"

Elizabeth: "In theory, yes. In practice, no. If you are in charge, you get recognized by the international community as the official government entitled to use the 'war power.'"

David: "Whereas just war theory would say the power test isn't demanding enough; there needs to be a moral test, too. The government must not simply be in charge but also *exercising its power in a morally fit way.*"

Nick: "Like how?"

David: "Most current thinkers would say 'in a way consistent with the individual human rights of the people it governs.' Governments exist to help fulfill the human rights of their people. Rights to life, liberty, . . .'"

Nick: ". . . and the pursuit of happiness!"

David: "And so a government is not legitimate, and doesn't deserve to be seen as having 'war powers,' if it violates its own people's rights."

Lori: "So, dictatorships aren't legitimate governments?"

David: "That's correct, they're not. They have power, but no legitimacy."

Lori: "Whereas realists would say their power gives them legitimacy."

David: "A mistake, in my view, because then they can't deal with the whole issue of humanitarian interventions, like in Rwanda or Darfur. When a government turns ferociously against its own people, and the citizens desperately need our help, what are we to do?"

Lori: "Don't the rights of political sovereignty and territorial integrity demand we keep out?"

David: "Here's how I see it: aggression, as we've discussed, is the key. And I define aggression as *the use of armed force in violation of the most basic rights of people to live on their own, in peace.* But aggression can be committed *either* externally *or* internally: 'external aggression' is rights-violating force directed at another country whereas 'internal aggression' is rights-violating force directed at one's own people. In either case, committing aggression causes a government to lose, or forfeit, or give up, its legitimacy and its right to govern. We then do no wrong when we resist such a government, or try to overthrow it, with force."

Nick: "Very well summarized."

Lori: "But do we really 'do no wrong'? *No* wrong? What if the weapons and tactics we use to overthrow that government strike the civilians in that country?"

David: "That's the separate issue of *jus in bello,* though."

Nick: "Well, then, let's talk about that."

Just then, Dr. Leung entered the room.

Chapter Six

Mortal Combat

Dr. Leung wore a surgical mask.

Dr. Leung: "Lori, may I talk to you in private, please?"

Lori: "That's OK, Doctor. Nick here is family, and I'd just have to tell him anyway. And these people are my new friends, so you can discuss it with me in front of them. I need company and support right now, anyway."

Dr. Leung: "We collected all the available data, debated all our medical options, and now we've mapped out our entire strategy, and scheduled out all the surgeries. We think Gil will need three separate surgeries, and the first one, on the high neck and spine, has just begun. Two further surgeries will be done on the back of the brain, and then the side, if this one is successful. We expect it will be. The second two are riskier, but must be done. Further reconstructive surgery—plastic surgery—on the front side of his face might have to be completed in the future, but that is not for us to do here. That's for back home. Gil has enough to do to make it through these three operations on the brain and spine."

Lori: "Thank you very much for the update, Doctor. I like to know the plan, and am glad to see he's got such a dedicated team. Can you give me a sense as to how long the procedures will take? And more on the risks please."

Dr. Leung: "I can only give you a general range for the timing. The vital thing, of course, is that the surgery be done well, and sometimes things take longer than what we expect, or else problems arise. For this first one, we're hoping for maybe a 10–12 hour time-frame, with three surgical teams rotating. Then, we'll look at the next procedure when we cross that bridge, but safe to say they will both be much longer, at least double the timing, probably."

Lori: "OK, Doctor, I won't keep you from the surgery. Thank you again, very much. Good luck with Gil."

Dr. Leung: "Thank you, Lori. He will be getting our very best surgical attention, and he's young and otherwise healthy. I will report back to you sometime tomorrow on the progress of the first procedure."

Lori: "Thank you."

After the doctor left, Nick said: "Well, its dinnertime. Let's find out what they serve for dinner, right? Everyone hungry?"

Nods all around.

Nick: "All right, then, let's get some calories in us and kill some time by talking more about war. By the time we're done, we'll have covered, and solved, it all!" He laughed.

The group—Nick, David, Lori, and Elizabeth—made a quick, brisk walk over to the mess hall. It was past the main dinner hour and so they were served heated-up leftovers from the main meal. Good, though; roast turkey with gravy, mashed potatoes, fresh carrots, and cooked corn. With three jars of pickles placed out as condiments. Elizabeth asked herself: "How many pickles do they think we'll eat?" Then the General pulled rank, and got some nice local German white wine—a sweet, almost sparkly Reisling—to wash it down.

The General—already pouring himself glass number two—began: "Thanks so much, Liz, for spearheading the *jus ad bellum* discussion. Fascinating stuff. Now I know that, as a military lawyer, you'll know more than me about *jus in bello*—the legality and justice of conduct in war—but I know a thing or two about this topic, and I'd like to give it a go."

Elizabeth: "Please, go ahead."

Nick: "OK, let me put on my lecturing cap, like the good professor here."

David chuckled: "Make sure the cap fits!"

Nick: "Ha! Well, you guys will let me know, right? OK, so the first thing to note is how the rules—contained in just war theory and international law—apply to a different crowd. Whereas the *ad bellum* rules apply to heads of state, and those with the 'war power,' the *in bello* rules apply to those who actually do the fighting; soldiers, and the officers who command them. The good ole' 'army, navy, air force, marines!' The rules regarding how to fight well."

Lori: "But it's silly, no? These are like boxing rules, or something?" She continued, mockingly: "'Here are the rules according to which we are going to try to kill each other, and destroy each other's social infrastructures.' It reminds me of the realist, who would say: 'Forget these rules. War is war and, once it's on, you fight to win.' Why fight according to rules?"

David: "Well, for several reasons. Most crucially, you don't want the war, and the destruction, to get completely out of control. You don't want it to escalate, say, to the point where civilians are getting targeted and slaughtered. The rules are designed to limit, control, and constrain the war and to prevent what they call 'total war.'"

Lori: "What's that?"

David: "Precisely the 'anything goes, nothing is out of bounds,' approach to war. I guess 'total,' in the sense of holding nothing back, everything becomes a target. A recipe for complete annihilation."

Lori: "I guess that's important."

David: "And why the realist is wrong again. The other big reason for having *jus in bello* rules is not just the desire to avoid the horrible consequences of total war, but also the sense that some ways of fighting are just wrong and morally corrupt, regardless of the consequences."

Lori: "Like what?"

David: "Well, I'm sure Nick is going to walk us through some of them. And it *is* quite like boxing rules, and that's really nothing to apologize for; there's going to be a fight, but that doesn't mean you let the fighters kick each other in the groin, or conceal weapons in their gloves, or keep punching after the bell has rung, or get a friend in the crowd to break a chair over the head of the other guy . . ."

Nick: ". . . or put some poison in the other guy's water bottle."

David: "Exactly, and so on. You get the point, there's fighting fairly, and there's fighting viciously. The *jus in bello* rules specify what fair fighting in wartime is. You know, it's sad that it has come to this but, as a matter of fact, it has; and so we try to restrain it and keep it as fair and humane as we can. You know, a number of the *jus in bello* rules actually have their origin in jousting tournaments, and sword fighting festivals, back in medieval Europe. Again, it is this sense of there's a fair way to fight and an unfair, wild, animalistic, unrestrained way of fighting."

Lori: "But we're talking *war* here, not a couple of aristocrats engaged in a personal duel with swords. The stakes are much higher."

David: "Not necessarily, not for the participants in such a duel, who knew they could be killed or seriously and permanently wounded, like losing the use of a hand, or an eye, and so forth."

Lori: "Wouldn't it actually be more humane simply to let everyone fight however they want, and have it over with?"

David: "I don't think so, because fighting unfairly *doesn't* stop the fighting sooner. It's the reverse actually. When you're been dealt an unfair blow, you don't surrender, do you? No, you get very angry, plot your revenge, and execute your own unfair blow. Unrestrained fighting is a recipe for escalation. There's no evidence whatsoever that it brings the war to an earlier end."

Nick: "I think that's been true in my experience on the battlefield, too. Makes psychological sense."

Lori: "I'm not so sure, though. What about Hiroshima? Wasn't that unrestrained fighting—and didn't it end the Second World War sooner than would've been the case?"

Nick: "Good counterpoint."

David: "Well, that brings up other issues, too. I think, before we can tackle Hiroshima, we should develop the other *in bello* principles which provide the full context for our evaluation of the use of the atomic bomb. Noncombatant immunity, especially. Nick?"

Nick: "OK, OK. But Hiroshima is definitely on the table, and we've got to return to it. OK, let's go back to the development. *Jus in bello* rules are part of just war theory and international law. Liz, which international treaties are the big ones here?"

Elizabeth: "The Hague Conventions, 1899–1907, and the Geneva Conventions, 1946 and 1977."

Nick: "Yes, yes, the Geneva Conventions. Everywhere, they're everywhere. Every soldier and officer must know about them. Every soldier and officer is held accountable to them and, after the war, if they've violated the relevant principles, they can be subject to war crimes prosecution, either within their own armed services . . ."

Elizabeth: ". . . by people like me."

Nick: ". . . hey, yeah, that slipped my mind. I guess I never should've volunteered to lead this discussion, right? What was I thinking?! Ha! I should've put you in the spotlight again."

Elizabeth: "Hey, you're the general, General."

Nick: "Now, now, none of that. The other kind of war crimes prosecution is by the international community. So, we know the point of *in bello*, to whom it applies, and where to find it in international law. Next up, the actual content of the *in bello* rules. The first and biggest, which David has already mentioned, is noncombatant immunity."

Elizabeth: "Easily the most frequently and strongly worded, *in bello* principle in international law."

Nick: "And the basic content of the principle is this; there are legitimate targets in wartime, and there are illegitimate targets in wartime . . ."

David: "And this is absolutely foundational to there being any *in bello* at all: some targets are off-limits. Otherwise, it would be a surrender to the realists and total war, with everything and everyone being a potential target. There needs to be a distinction, or discrimination, between permissible and impermissible targets."

Nick: "And only the permissible ones may be targeted with force. Yes, I remember my War College instructor calling the rule 'Discrimination *and* Non-Combatant Immunity.'"

Elizabeth: "They really are bundled together, logically. To avoid total war, some targets must be off-limits; so discriminate between the legitimate and illegitimate targets; and then only fire at the legitimate targets."

Nick: "And the legitimate targets are military ones, and the illegitimate ones are civilians. Civilians are defined everywhere in international law as 'noncombatants' and so, you put it all together and you arrive at the conclusion that noncombatants, that is, civilians, may not be targeted at all with military force."

Lori: "Civilians may not be killed in war? But that rule gets violated all the time! So much for the effectiveness of international law."

Elizabeth: "We need to make a careful distinction here, actually. It's going to sound funny at first. It is actually not illegal to kill civilians in wartime; what is illegal is to *intend to kill* civilians in wartime, notably by aiming at them directly and deliberately."

Nick: "The notion is this, *Lori:* mistakes get made. War is chaotic, loud, smelly, and violent. Sometimes civilians happen to be close to the military target at which you're aiming. Even worse, mistakes in intelligence gathering and data analysis occur, and pilots in planes, or soldiers in tanks, get instructed to blow up a target—like a building—which turns out to have been civilian in use. Again, mistakes happen. The fog of war. So, *it's not a war crime if you mistakenly kill some civilians.*"

Elizabeth: "Such civilians are defined as 'collateral damage.' Literally, 'damage on the side,' accidental casualties which you did not intend and, crucially, which you did not deliberately target with killing force."

Nick: "Unlike Calley, for example."

Lori: "Calley?"

Nick: "One of the most famous war crime cases from the Vietnam War; Calley was the leader of a U.S. unit raiding the Vietnamese village of My Lai. Now, as you know, Lori, the Vietnamese fought the foreign forces in Vietnam—the French, and then the Americans—in a guerrilla-style campaign. This is to say that, whenever possible the Viet Cong refused ever to come out and fight the Americans openly in what we call 'a set-piece battle.' A set-piece battle is when you've got one army clearly and openly fighting against another. The classic example would be the old European battles, in the open fields, with the two sides charging against each other. Well, anyway, you can see how the Viet Cong wouldn't want to fight that way . . ."

Lori: ". . . because superior American technology and firepower would blow them away . . ."

Nick: "Correct. So, they would do ambush-style attacks on the Americans, hiding in the jungle. Then, after the battle, the Viet Cong would hide in small villages, take off their war garb, and try to blend in with the civilian population."

Lori: "Clever."

Nick: "And brutally frustrating for guys, like me, required not to target civilians, yet still to try to defeat the Viet Cong! Just pure hell. Anyway, Calley's unit eventually had enough of being subject to this style of fighting, and they just snapped. When the villagers of My Lai refused to identify the Viet Cong soldiers in their midst, Calley ordered the massacre of the villagers. Deliberate civilian targeting."

Lori: "But wouldn't some of the targets have been soldiers?"

Nick: "Yes, but others were old men, women, and children. Executed. Deliberately. Major war crime."

Lori: "But wasn't it wrong for civilians—the villagers—not to identify the soldiers in their midst? You know, I'm actually having problems with the sense of this so-called 'all-important,' or 'ever-present,' rule. Don't some civilians *deserve* to be targeted? Like these Vietnamese civilians who clearly hid and supported the Viet Cong? Or what about civilians who vote in favor of a war, like those Americans who voted for Bush? And, now that I think of it, what about the other way round? Do all soldiers deserve to be targets? What about all the poor, illiterate farm boys around the world who've been conscripted and forced to fight by their local dictator? Is it fair that they can be killed, while rich civilian businessmen give money to support that local dictator, as long as he leaves them alone to conduct their corrupt business practices?"

Elizabeth: "Well, these are difficult issues, especially philosophically. But I like international law's way of handling it. The law says: 'Look, we've got no way of knowing which civilians have voted for the war and which haven't. We have no way of peering inside the skulls of everyone on the enemy side to see who supports the war and who doesn't. All we can reliably go on is external conduct and behavior. Soldiers have weapons whereas civilians don't. Whether a soldier is conscripted or not doesn't change the fact that he's in the other trench, or whatever, and firing at you. It might make it tragic, in some cosmic sense, if he gets shot by you even if he disapproved of the war, but was forced to fight in it. But that doesn't change the crucial fact that, from your point of view, you didn't, and couldn't, know his internal attitude. And he was shooting at you, too. It's his *status* as a soldier, and his *being a source of harm*, that makes him a legitimate target for you. Whereas civilians aren't soldiers—it's not their business or training to deploy armed force—and they are not a source of violence directed at you. Thus, you may not target them.'"

David: "An excellent explanation, Liz. But we all know there are grey areas and difficult problems here. Let's put them on the table. What of Osama Bin Laden's argument that civilians *do* cause harm, even if they don't carry weapons? He refers to taxes. The military is directly enabled by the taxpayer to do the harm it does and so, he infamously concludes, everyone—man, woman, and child, soldier or civilian—on the enemy side is a legitimate target."

Lori: "Doesn't that actually make sense, though? Wouldn't that make the fighting much easier?"

Nick snorted, then replied: "Yes, it would sure make the fighting much easier. I wouldn't have to care who I'm shooting at, or be at all diligent in how I ordered my men to fight. It would be total war, Lori!"

David: "Indeed it would. Both points count against Bin Laden; the result would be total war, which we've already criticized. And it would make the fighting too easy for the soldiers, whereas we want them to maintain their fighting discipline and to focus their efforts only at the source of harm, which is the military-industrial-political complex, aimed against them. Plus, we can't forget Bin Laden's own huge personal bias in favor of making this argument; it justifies terrorism. If everyone is a legitimate target, then it's alright to fly planes into buildings, and to blow up trains, and to use suicide bombers to detonate themselves inside busy churches, mosques, and marketplaces."

Elizabeth: "Plus, it still remains wrong to consider children as legitimate targets, even if you buy the taxation argument, as kids don't pay taxes."

Lori: "But they're potential future taxpayers."

Nick: "Just like my current friend and ally is a potential future enemy, so why don't I just shoot him in the head right now? Like America should drop a nuke on Canada tomorrow, right? That way lies madness, Lori. And, with kids, as we all know, they don't even have the developed moral character to be considered guilty of anything. Thus, to do anything in war which will intentionally result in their death is murder. It is the intentional killing of an innocent person."

David: "So the only defensible conclusion is that some groups of people morally must be off-limits in wartime."

Elizabeth: "And so we've made that the law."

David: "But, to continue with the difficulties and problems, let's consider the civilian workers in a weapons manufacturing plant. On the one hand, they are civilians, apparently, and off-limits. On the other, the factory itself would seem fair game, as it is part of the military-industrial complex."

Lori: "So, what's the solution? What's the relevant rule?"

David: "The rule is that such workers may be fairly targeted while working, but not when they are off work and relaxing at home."

Lori burst out laughing: "You've got to be kidding me!"

Elizabeth: "No, he's not, Lori. Think about it, and remember that, in a way, it has less to do with the workers as individual human beings and more to do with the permissibility, or not, of general targets within which that individual human being might be located. We've established that a weapons factory is a legitimate target, right? Clearly a source of harm. So, when the workers are there, they are part of that harm, part of that military-industrial-political complex. Thus, they are fair game, and wearing a bull's eye on their forehead while they are engaged in such activity. If they don't like that, they can secure another job. But when these workers leave work, they go back home to a residential area, away from the factory, and in that residential area are a bunch of civilians, including many who have nothing whatsoever to do with 'the war machine.' Think of the very old and the very young, especially. Plus, while the workers are at home, they are relaxing and not engaging in harm. So, while at work, they are a target; while at home, not."

Lori: "But what about things used by *both* the military and civilians; like roads and bridges?"

David: "And electricity-generating stations, and radio and TV airwaves, and water-cleaning plants, and hospitals . . ."

Nick: ". . . and farms which give food to everyone."

Elizabeth: "These are all called 'dual-use targets' and international law strictly forbids their targeting in wartime. But, in practice, it must be said that many militaries do go after them or, at least, some of them. Communication facilities— radio and TV stations and transportation networks in particular: roads, railways, airports, and bridges. The rationale given is that, *while the war is on*, the enemy makes exclusively military use of such things, and so they are fair targets."

Lori: "That can't always be true, though, especially of roads."

Nick: "No, it is not always true, but in reality, it often is. When a war is on, few civilians want to get in the way, or risk their lives driving across bridges. And the government just takes over the airways to fill it with pro-war propaganda. So, taking these things out isn't a big deal."

David: "I agree. More disturbing are things like blowing up water-filtration and cleaning plants, which America did during the Persian Gulf War in 1991. Targeting a vital need, like clean drinking water, is wrong. I believe that was a mistake, though. What wasn't a mistake, I hear, was targeting Serbia's electricity grid during 'the shock and awe' campaign there during the 1999 NATO bombing raids of Kosovo."

Lori: "Uh, what was that about again?"

David: "Kosovo, a province of Serbia, wanted to separate. Kosovars are ethnically different from Serbians. Serbia resisted Kosovar independence with military force, and NATO punished Serbia with a retaliatory bombing raid which caused the Serbian government to fall."

Nick: "And David's right, we deliberately damaged and then took over the electric grid, as a demonstration of power, and to pressure the people to overthrow the government."

Lori: "And it worked?"

David: "Indeed. But that doesn't justify it, especially when there were other means. During 'Iraq 2,' we did the same thing—the same 'shock and awe' display of overwhelming force, and many dual-use targets were hit: water, electricity, hospitals, communication towers, and so on. Some neighborhoods in Baghdad had their basic infrastructures completely ruined for a year, or even more."

Lori: "Wow. Is that commonly known? So, it's not OK to target residential areas, or dual-use targets. Or, I guess, farms and schools and things like that. What *is* permissible to target?"

Elizabeth: "Well, the general principle is the only permissible targets are those clearly connected to 'the war-machine': the military-industrial-political complex which runs and operates the war effort. Everything else is an illegitimate target, including all dual-use targets."

Nick: "Generally, illegitimate targets have little military value, anyway. Shooting up kids in schools, for example, doesn't help you win the war."

David: "True, unless the purpose is to terrorize and demoralize an entire population."

Elizabeth: "And that's why we have the rule in international law."

Lori: "But how do you—like, literally, *you* guys in the military—how do you assess and distinguish between military and civilian targets?"

Nick: "It's called 'due care.' Civilians are owed due care; all reasonable, diligent, professional efforts that they not wind up casualties of war. You make that effort part of the entire structure of the military, right from the top straight down to the bottom. At the very top, those who craft overall war strategy must create battle plans which minimize the risk of civilian casualties. Your military needs intelligence and espionage, needs data-gathering and data-analysis capabilities, to do this in a compelling and informed way. So, you've got to do that, too. You've got to train your soldiers in these principles, and write them into international law. You've got to build weapons—like laser-guided, satellite-guided, and radar-guided ones—which are less likely to kill the wrong people. And the soldiers and commanders on the ground, during battle, must do their very best to retain a level head and not run off and do the sort of thing which Calley did. Excellent training, and good combat experience, are the key here."

Lori: "Good lord. That's all just so much effort, and it's so demanding and costly. I keep coming back, in my mind, to the realist point of just fighting any which way you want. War is war. I mean, why bother with it all, especially when so many mistakes get made in spite of all these excellent efforts."

Nick: "Two reasons: one, moral and the other, strategic. And they've both been said, Lori, so pay attention, please. The moral reason—and it's huge—is that *you don't want to kill people who don't deserve to be killed.* They have a right to life and, my God, I assure you that the very last thing I want on my hands, as a soldier, is the blood of an innocent person. I'm no criminal or common thug. My job is to defend my nation. And the strategic one—and surely the realist can appreciate this—is that, as I said, *often illegitimate targets are useless.* They do not advance war aims at all. You want the bombs to fall on factories, not apartment buildings. You want roadside explosives to take out tanks, not cars. You want bullets inside the guts of some soldiers, not some shoppers."

Lori: "OK, I get it. But didn't I read somewhere that modern wars feature more civilian than military casualties?"

Nick: "True. I think all of them. Every one since WWI."

Lori: "Then isn't there something going on with that? All these efforts get made, yet not only do civilians wind up dead, they wind up dead in much greater numbers than the soldiers."

David: "Well, that is a disturbing truth, and obviously there needs to be more due diligence and care regarding civilian casualties. But, surely, that's no reason to get rid of the principle, the fact that it is routinely violated, right? Should we get rid of property laws because people steal?"

Lori: "OK, good analogy. But I've come up with a really big problem here. I thought of it when you three were talking about children being innocent of war."

Nick: "Shoot."

Lori: "The problem of the child soldier. I've been reading into this. Apparently, not unknown in history and a recent, disturbing trend and reality, especially in the wars of sub-Saharan Africa."

Elizabeth: "Correct. I think the experts say the modern phenomenon began with the civil war in Mozambique, in the early '80s. The rebels there kidnapped boys from their villages, trained them, brainwashed them, gave them weapons ranging from machetes to machine guns, got them ready for battle—treating it like a game, of course, and often feeding them drugs—and then unleashed them on the government soldiers. Unspeakably horrible."

Nick: "Any bastard officer who did such a thing should be shot on sight."

Lori: "But my question and challenge has to do with legitimate targets: is a child soldier a legitimate target? On the one hand, he's a child. I hear some of them are as young as eight or nine, obviously before the age of reason and moral responsibility. On the other, he's a soldier and shooting at you."

Elizabeth: "It is a very disturbing scenario, Lori, and a real one throughout Africa and southeast Asia, too. International law has two answers here, and I think one is good and the other is bad. The good answer—'good' only in the sense of being correct, not happy—is that a child soldier is a legitimate target because, as you noted, he's shooting at you. He's a soldier, armed and dangerous, regardless of age. The bad answer is that international law doesn't, as yet—not in my view, anyway—have anything near the kind of penalties and punishments which need to be put in place to prosecute the adults who use child soldiers. They must be put on war crimes trials and punished for this."

Nick: "Shot on sight, I tell you."

David: "Plus, there needs to be more efforts—like those by the UN and international Non-Governmental Organizations like 'Save the Children'—to rehabilitate former child soldiers so that they are not utterly ruined for life. I mean, can any of us imagine what it would be like, when we were just nine, to have killed another person, and to have been shot at, and been in real wartime battle? It completely boggles the mind. Those responsible for such atrocities must be brought to justice."

No one disagreed.

Lori: "What are the other *jus in bello* rules?"

Nick: "Well, there's another one derivative from this basic, three-fold package of discrimination, due care, and noncombatant immunity. And that's benevolent quarantine for prisoners of war. If enemy soldiers surrender, and lay themselves and their weapons down, they are not to be shot or killed, since they are no longer a source of harm once they have surrendered, right? Such fellas are to be taken into custody, taken away from the front lines of the battlefield, given basic food and shelter, and then swapped for your own captured soldiers at war's end."

Elizabeth: "The key concept is 'benevolent' in this benevolent quarantine rule. You can question captured soldiers, but you cannot torture them, or force them to fight against their own side, or starve them, and so on. They must be kept dry and fed, with their basic hygiene—soap, shampoo, toothbrush and paste, and so forth—provided. No torture. The Geneva Conventions are all about permissible and impermissible ways to treat prisoners of war, like those held in large-scale concentration camps and the Japanese POW war camps, and they were drafted and passed in 1946, just after the Allies discovered what went on in the Nazi concentration camps during World War II."

Lori: "But this is still hugely relevant today. I think of the whole issue of torturing terrorists. So, America was violating international law when it used torture during the questioning of terrorists, and terrorist suspects, like in the Abu Ghraib prison in Iraq, and at Gitmo in Cuba?"

Elizabeth: "Well, the answer is not totally clear. The Bush Administration admitted that torture, or torture-like, questioning sessions were taking place. For instance, the use of water-boarding during questioning. This is when your head is repeatedly dunked underwater and held there to 'soften you up' during questioning. Like bobbing for apples—*only you are the apple!* The theory is to make you admit things, and share information about terrorist groups and future planned attacks, that you wouldn't otherwise admit if you were warm and fed and comfortable."

Lori: "But, that's torture, right?"

Elizabeth: "Well, yes and no. Yes, the activity itself is prohibited by the Geneva Conventions. As is the use of chronic sleep deprivation and the use of animals— like attack dogs—as a tool of intimidation during questioning. But the Geneva Conventions—as I understand them—only prohibit doing such things against captured soldiers, or civilians, and the Bush Administration's position was that terrorists and terrorist suspects are neither, and are part of a third group—of outlaws— and so they don't deserve the protections established by the Geneva Conventions. Technically, they actually had a legal argument there, in spite of all the negative publicity about the thing."

Lori: "But not morally. I mean, c'mon! No one was worse than the Nazis, right? If it was wrong to torture captured Nazi soldiers, then it's wrong to torture terrorists, much less mere suspects! Or, lemme think, we don't let prison guards in domestic society torture even the very worst convicts in jail—rapists, murderers— so how can we allow such behavior at the international level?"

Nick: "The thinking is that we need this information to prevent another 9/11."

David: "But, Nick, if you read Daniel Mannix's *A History of Torture*, you'll see that torture is a very dubious device for gathering correct information. The whole notion of pain 'breaking down the barriers' which stand in the way of truth isn't right. It seems that victims of torture will say anything—absolutely anything, whether true or false—to get the torture to stop. The whole objective of the victim, apparently, becomes to please the torturer so that they'll stop. So, you confess to anything. You make things up; it's all just an attempt to get the pain to stop."

Nick: "Your point?"

David: "The point, and it's crucial, is this; that the very rationale for the torture is incorrect! Torture *doesn't* get good information, plus it is horribly wrong, owing to the pain and suffering it inflicts."

Lori: "But maybe that's the *real* point behind the torture; not so much to get good anti-terrorist information as it is to send out a clear and very intimidating message to those involved with terrorism, or contemplating such: 'if we get our hands on you, this is what we'll do to you.'"

Nick: "And that itself can have a deterrent effect, no?"

David: "Well, maybe. It's easy for the authorities to say so, isn't it? We'll never know. Myself, I believe it is not the American way and am deeply uncomfortable with the idea that Americans are, or were, doing this. It is a violation of human rights in the name of national security."

Nick: "But it's the tough business which must be done, in wartime, to win. That's why, with respect, you teach at West Point whereas I command in the field. It's a nasty world."

Lori: "I'm with David, and thank God Gil wasn't involved in such activities. I shudder to think about the kind of people who are, whether American or not . . ." She shook her head. "OK, more *jus in bello* rules."

Nick: "OK. Proportionality, again. Here it is a requirement of proportionality, or balance, between the tactical objectives of the battle (or maneuver), and the force needed to achieve it. But this is just sound strategy, in my view; only use that amount of force needed to secure your objective. It's an economy of force principle; don't squash a squirrel with a tank."

Lori: "But then, if it's consistent with military strategy, does it need to be made part of military ethics or law?"

Nick: "Yes, because sometimes during the heat of battle officers and/or soldiers make bad choices, and deploy wasted force. It's quite common to do so out of anger: you suffered something, so you want to make the enemy suffer that times a hundred. Got to rein in those tendencies."

David: "A commonly agreed upon example of disproportionate force on the battlefield, Lori, came during the closing days of 'Iraq 1.' The Iraqi army was in full retreat, along the one major highway connecting Kuwait and Iraq. So many of them, at one time, created a massive traffic jam. When the American military discovered them, they just pummeled them."

Nick: "I was actually among those Americans."

David: "Oh, my! Well, uh, Walzer argues that the killing there, along what came to be called 'The Highway of Death' was too easy. 'A turkey-shoot,' he labeled it."

Nick: "But the Iraqi's didn't surrender, so they were legitimate."

David: "I don't deny that. It wasn't a violation of discrimination, or benevolent quarantine, but rather of proportionality."

Nick: "Easy to say that when you weren't there, or after the fact. On site, you are just glad for the victory—and the easier, the better. Now, moving on to another rule; no means '*mala in se*' are to be employed. Liz, what's the exact legal Latin translation?"

Elizabeth: "It is prohibited to use weapons or tactics defined as being 'bad-in-themselves,' or 'evil-in-themselves.' These are things ruled out, not just because of the pain and suffering they cause, but moreover, because there is just something intrinsically horrible about them."

Lori: "Like what?"

Elizabeth: "Like forcing POWs to fight against their own side."

Lori: "That is low."

Elizabeth: "Another example is the use of rape in wartime."

Lori: "Disgusting!"

Elizabeth: "It happens much more than people talk about. Historically, rape has been used by armies to drive a population off a territory—create horrible memories there, which make them want to leave—and also to reward their troops. Even countries which we often consider 'good guys' have done this in the past: 'win control of the town, boys, and you can have your way with the local women.'"

Lori: "Wretched. They should all be castrated. 'Evil-in-itself,' indeed. Only a man would think of using rape as a tool of war."

Nick: "Uh, I'll let that one slide, in the interest of moving on. There are two more rules, as far as I can recall. One, no reprisals. A reprisal is this; the enemy violates one rule of *jus in bello*, and you violate *jus in bello*, too, in retaliation and in order to punish or chasten the enemy back into a state of mind where they will obey the rules once more. Say, if you were to drop a bomb deliberately on the enemy's civilian population in retaliation for the enemy previously having done the same thing. This happened near the close of WWII in 1945, when Allied forces leveled the city of Dresden in Germany through firebombing. This was done in retaliation for Hitler's previous targeting of Britain's civilians in London and Coventry. This is strictly forbidden: the enemy's violation of the rules does *not* give you justification, or an excuse, for violating the rules yourself. The rules are the rules are the rules."

Lori: "And the last one?"

Nick: "No use of prohibited weapons. There are many such things, and many, many international law treaties specifying them, and outlawing them."

Lori: "Like which ones?"

Nick: "Anything dipped in, or which makes use of, poison or pestilence. Probably the oldest rule here. Think back to medieval times: if a duelist dipped his sword in poison and then cut the opponent, this was considered complete cowardice, evil, and treachery."

Lori: "There's a scene in Shakespeare's *Hamlet* about that. Near the end."

Nick: "Well, there you go."

Lori: "There must be more."

Nick: "You better believe it: land-mines; dumb-dumb bullets which expand on contact to do maximum damage; bullets which contain glass shards . . ."

Elizabeth: "There's a treaty prohibiting weapons, or tactics, which substantially alter the natural environment. People thought Saddam should've been up on charges, under that treaty, for when he set all those Kuwaiti oil wells on fire during the Iraqi retreat from Kuwait in 1991."

Nick: "And then, of course, there's the weapons of mass destruction."

Lori: "Like nukes. Finally, back to Hiroshima."

Elizabeth: "But nuclear weapons are not forbidden by international law; only biological and chemical weapons are."

Lori: "What?! Prohibit fancy bullets, but leave nukes to be legal? What the hell is up with that?"

Nick: "Well, the most powerful countries in the world are all nuclear powers. Their having this 'nuclear edge' gives them a lot of leverage, and strikes fear into their opponents. So, they don't want to lose that leverage by having to give up their nukes, and so they have all conspired to block the passage of any treaty which would ban nuclear weapons."

Lori: "But they've agreed to such treaties on biological and chemical weapons. What's the difference, by the way?"

Nick: "The difference is that biological weapons release a living organism—a bacteria or virus—which attacks or eats vital human tissue. Imagine a bomb which released the Ebola virus, or bird flu. A chemical weapon by contrast, doesn't release a living organism. It releases an inert substance—usually a gas, like mustard gas or saran nerve gas—which usually causes suffocation. The gas blocks out breathable air and people choke to death from lack of oxygen."

Lori: "So, banned?"

Elizabeth: "Banned from use, not from possession. The major powers all keep massive stockpiles of these things."

Nick: "You know, 'just in case' . . ."

Lori: "But not nukes. Let's bring Hiroshima back on the table. America, having won the war in Europe in 1945, turns around and drops two atomic bombs on

Japan in August 1945. Massive casualties, the majority civilian. Terrible. And Truman had the gall to say that, after having ordered the use of the nukes, he never lost one moment's worth of sleep over it."

David: "Elizabeth Anscombe, upon finding out that Truman was to be awarded an honorary doctorate from Oxford—her university—wrote and published an essay entitled "War and Murder." In it she blasted the U.S. president for using the bomb, labeling it the murder of innocent Japanese civilians."

Nick: "Yes, but again with respect to you academics, that's easy for her—years later—to say. She didn't bear the burden of wartime decision-making, whereas Truman did."

David: "But the atomic bombing did kill well over 100,000 civilians—innocents, no? What was the combined casualty rate from Hiroshima and Nagasaki. I think it was just under 300,000. Now, some of these would've been military, but lots of them—the majority, right?—would've been civilian. And then you throw in those who survived, yet contracted radiation poisoning, and it's all just a disgusting mess. Appalling. And that the president of a democracy ordered such a thing, I have to say, makes me shudder."

Lori: "Me, too. That's what happens in war, this race to the gutter, this progressive lowering of standards of restraint and conduct until a point where such things get ordered."

Nick: "Neither of you are thinking clearly. This is all just such easy criticism of Truman. Face the facts; Truman was forced to use the atomic bombs, for two reasons: 1) Japan was beaten, but irrationally and unmorally refused to surrender because of their samurai code of honor; and 2) if the United States had to invade Japan to force the final admission of defeat, the predicted casualties were 1 million people. That's in total: military and civilian, both Japanese and American. One million dead, versus the 300,000 dead which the nukes produced. And the invasion would've generated civilian casualties on a huge scale, too. So that was Truman's choice, forced upon him by the Japanese refusing to surrender: drop the bomb, or invade. Either way, there would be massive civilian casualties. If you drop the bomb, though, that just takes two planes and two bombs whereas, if you invade, it will take massive preparation in terms of time, money, and personnel."

David: "But the appeal to financial cost, and the logistical difficulty of mounting the invasion, can't outweigh the moral fact of *knowing* that your decision will murder civilians."

Nick: "But these civilians, and quite probably more, would've died during the invasion anyway. If you ask me, comparing a casualty total of 1 million to 300,000 is a no-brainer, and Truman made the right choice."

Lori: "But the 300,000 figure is real—actual dead people from the actual dropping of the real bombs—whereas the 1 million figure is speculative and made-up, based on guesses of what might have happened, yet didn't, and so who are we to say they're wrong?"

Nick: "Well, who are we, indeed? The very best planners in U.S. policy and military circles came up with that figure. It's not a mere guess, but a calculated estimate."

Lori: "But you see, Nick, how these planners had a vested interest in inflating that number, making it as big as they could reasonably get away with. They wanted the simplest solution for themselves—drop the bomb."

Nick: "Well, what's wrong with that? The decision was simpler, it cost less, it did in fact force the Japanese to surrender, and it produced fewer casualties."

Lori: "That's my whole point! We don't know whether it produced fewer casualties."

Nick: "Well, I believe it did."

Lori: "But believing and knowing are two different things."

Nick: "But you can't deny the atomic bomb forced the Japanese to surrender."

Lori: "No, I suppose I can't."

Nick: "There you go."

David: "But the fact of victory doesn't justify doing anything you want to achieve it. The ends don't justify the means."

Nick: "Well, in war, the ends count for a lot. Like who wins and who loses."

David: "The results are important, I grant you that. But, again, it matters how these results have been produced. People shouldn't pride themselves on creating good results through bad means."

Nick: "But the means weren't bad, as they produced fewer casualties than the only other alternative."

David: "So you say, on the casualty rate—and that's Lori's point—whereas let me stress that what was bad about the means was the deliberate, foreseen, and intentional killing of civilians on a massive scale. You simply can't pretend there's nothing wrong with that."

Nick: "There is something wrong with it, but I just don't think Truman had any better option."

David, yawning: "Well, we'll have to agree to disagree on that."

Elizabeth: "Yes, David's right: I mean, in terms of yawning, it's getting quite late for us all."

Lori: "Yes, a great set of conversations today. Maybe we can meet at breakfast tomorrow, unless I'm called away to attend to Gil? At 9 a.m. in the mess hall?"

They all agreed, found their rooms, and went to bed.

Chapter Seven

War's Aftermath

In the mess hall, at breakfast, were Elizabeth, Nick, and David. On the menu: undercooked scrambled eggs, overcooked breakfast sausage, buttered toast, cereal for those who wanted it, and then enough coffee and milk to sink a ship. (David asked himself: "how can someone ruin toast?") But at least the coffee was tasty. Everyone was tired and groggy, and conversation was spare and silent.

When Lori walked up to their table, she was obviously fidgety, and filled with stress and tension. Her eyes were focused and intense; her straight hair looked wild and wispy.

She announced: "Dr Leung just told me that the first surgery went very poorly, and that Gil is now fighting for his life. He's starting to bleed internally in the brain, or something. Oh God, Nick . . ."

Nick stood up, took her into his arms and gave her a long, warm, wordless, comforting hug. Everyone else looked on. After a moment, Nick said: "So, what's the next step?"

Lori: "Gil's back to the operating room, straight away. No time to recover from the first, as he started to crash in the middle of it. They brought in a new surgical team, and Dr. Leung said that now they have to be very aggressive and essentially try to pack all the surgeries into one marathon session, while trying to save his life at the same time, I mean . . . now that I think of it . . . that must essentially mean he's dying, right? It's the last chance to save him, so they are just going for broke . . ." She started to sob violently.

Nick asked David to bring Lori a breakfast plate. When David returned several minutes later with two plates, each filled with food, Lori's sobbing had subsided.

Lori: "In the meantime, I'm powerless. I can only hope for the best."

Elizabeth: "Perhaps David and I should leave you two alone. Is there anything we can do for you, or your family? Should anyone be called?"

Lori: "That's very kind. But I think I'll wait to see if this last-ditch surgery will pan out. If it doesn't, I'll have lots of time—too much time—to make all the calls and arrangements."

Nick: "And I'll help with that. The army has people on staff who can assist with such things, too."

Lori: "Whoop-dee-do. Lucky me; they kill my husband, then are at least polite enough to arrange a hearse."

Nick: "Please don't start, Lori. Let's focus our thoughts and good energies on Gil."

Lori: "No, I want to talk, to continue our conversation. I won't attack the Army. I promise. But, more than ever, I want to talk about Iraq and what's going on there. To make sense of it—more fully process it—because Lord knows I don't want Gil to have died in vain."

Elizabeth: "Well, if you're sure you're up for it, Lori. I know exactly where we can begin. I was going to bring this up anyway, as it's a logical continuation of where we left off yesterday."

Lori: "Please, let's proceed. The doctors know where to find me."

Elizabeth: "Well, I was actually doing some Google searching on my laptop last night before bed and, and wouldn't you know it, discovered that we have here among us one of the world's most cited authorities on *jus post bellum*, or justice after war: Professor David Pearson of the West Point Military Academy."

Nick: "You're *that* David Pearson? I've read your stuff. It's very good . . . controversial, though . . . I'll be damned . . . I'm sorry that I never put two-and-two together. I've actually used some of your material when instructing my officers."

Elizabeth: "And I've used some of your material in my legal papers on war crime trials and broader postwar settlement processes."

David: "Hey, hey . . . hold on everyone. Not so much praise—and certainly not so early in the morning before I have a chance to jot it all down! It's all very kind of you, thank you very much, but I can't live up to that billing. I'm a professor— a just war theorist—and I've tried to put together some interesting thoughts on postwar justice in particular. My specialty is to advocate for the integration of human rights into postwar reconstruction plans. I've received some attention for it, and am grateful. But that's all."

Lori: "Still, it must be nice to know your ideas are being heard and discussed."

David: "Absolutely. Better, by far, than dwelling in oblivion!"

Lori: "So please, discuss them with us. Give us a summary lecture on *jus post bellum*, and we'll ask questions . . ."

Nick interrupted: "And demand answers!"

They all laughed, and appreciated the tension release.

Lori: "But please, can we focus on postwar reconstruction in Afghanistan and Iraq? The whole 'war on terror' business? For the sake of paying respect to what Gil has been involved with while in the Army."

David: "Sure, Lori, it'll be my pleasure. Now, everyone, I warn you; I can go on forever about this subject. So feel free to walk away when you can't stand it anymore."

Nick: "We will: no problem."

David: "OK, then, suppose this: you're the American president. Your military has just defeated some other country in war. Let's assume, for now, that you were on the just side during the war. In other words, congratulations, you've won. But now that you've won, what do you do?"

Nick: "Celebrate?"

David: "Yeah, yeah, sure. But after that, really, what would you do? You're holding the bag, now. Millions of ordinary people in that defeated society are looking *to you* to see what you'll do. It will have a massive impact on their lives, and their children's futures. It's an enormous responsibility. So, I put it to you again, what would you do?"

Nick: "Hey, you tell us, you're the supposed expert here."

David: "OK, there are two general policies which winners tend to follow. I call the first one the policy of revenge, and the other, the policy of rehabilitation. The first one is more cautious and conservative; the second more aggressive and risky. At least in the short-term; over the long-term it may well be the other way around."

Nick: "You've lost me already."

David: "Sorry, too theoretical to start with. Think of it this way, a sort of analogy. There's been a bad guy—a criminal—running around. The cops finally caught up with him and have taken him out of circulation. He then went to trial and got convicted. Now, what should society do with him?"

Nick: "Punish him, of course. And throw the jerk in jail. Fine him for the cost of his room-and-board, while you're at it."

Lori: "I disagree. I think punishment backfires, and just makes the criminal harder, and condemns him to an entire life of crime. I favor helping the criminal become a better person and get some decent job skills, so he needn't turn to crime for money and so he can then one day become a productive member of society."

David: "And that's essentially the difference between the two postwar paradigms, or policies, of revenge and rehabilitation. The revenge policy is all about punishing the defeated society, usually because that country is believed guilty for having started the war. The rehabilitation policy, by contrast, does not focus on postwar judgments about guilt for the war's beginning; the focus is on what's needed to make the defeated nation a better one in the future; a more peaceful, prosperous society."

Lori: "Why would anyone support the revenge policy? Just the word 'revenge' alone tells you it's no good, or not as good as the other. The word 'rehabilitation'—like literally, making better—is so much more positive."

Nick: "But that's just branding, Lori, that's just the choice of names which David has created. And, having read his stuff, I know that his choice of words may be influenced by the fact he favors the rehabilitation policy over the revenge policy."

David: "That I cannot deny."

Nick: "But I can tell you one reason why people might favor the revenge policy; they believe it is demanded by justice. So, I don't like the word 'revenge,' it's biased. I would describe the different policies, as I know them so far, as conservative versus aggressive. The revenge policy is conservative in that less is done, right? You don't have the big social changes, or postwar reconstruction changes. You don't engage in widespread reengineering of the defeated society. You focus on punishing it for having started the war, and then you leave the rest to the local government."

Elizabeth: "And you can see how others would pick up on that last point, and refer to Afghanistan and Iraq—and all the difficulties and challenges going on there—and say this: look, the rehabilitation policy is too costly and demanding, too risky and difficult . . ."

Nick interrupted: ". . . and we don't owe it to them to reconstruct their societies."

Elizabeth: ". . . and so let's not do that anymore. Let's just go back to the older, conservative way of punishing the country, and then leave and get out. That's all we're responsible for."

Lori: "But what exactly do you mean by 'punishing the country'? How does one punish a whole society?"

David: "Usually, the revenge or punishment paradigm will include the following elements: get the defeated society to admit it wrongly caused the war, and to apologize publicly for that fact; if the defeated aggressor gained anything from wrongly causing the war—say it gained territory by launching an invasion—then it must give up those gains, and restore them to whomever it took them from . . ."

Elizabeth: "But it can't simply give up its wrongful gains. It must additionally lose something further, so it feels punished and will not repeat the aggression in the future."

Nick: "Right, *it must be made worse off after the war* than it was before, so that it understands the wrongness of war, and won't be as tempted to go to war in the future. It's just like with disciplining a child, you can't just make it give up what it wrongly took, they must give it back; apologize; and then lose something further of their own—like dessert after dinner—so they are punished and will think twice about the consequences before being tempted to do the same in the future."

Lori: "I thought the doctrine of punishment as deterrence has been shown false. In domestic society, with criminals, the rate of repeat offending is very high, so punishment doesn't seem to prevent them, or deter them, from breaking the law in the future."

Elizabeth: "But that might just mean that it is not a precise analogy between individual criminals and a whole society defeated in war."

Lori: "Granted. But, look, if you made that society worse off after the war, this will make it harder for them to reconstruct, no?"

David: "Precisely."

Lori: "Plus, there's the issue of punishing people who don't deserve to be punished. I think back to our *jus in bello* discussion of discrimination and noncombatant immunity: why should civilians in that society suffer, and be made worse off, when it was a completely unaccountable regime, which they couldn't influence or control, which probably launched the aggressive war in the first place?"

David: "Again, a great and important point. But even conservative supporters of the revenge principle do respect discrimination in one kind of way; they do distinguish between what those truly responsible for the war deserve versus the broader civilian population. And they actually tend to be split on whether to punish civilians at all."

Lori: "OK, so what do those responsible for the war deserve?"

David: "I guess this would be the third element in the revenge policy: leaders of the defeated aggressor need to be removed from power, so they can't do the same thing again, and then to be put on trial for war crimes charges. No civilians deserve any of that, as they weren't responsible for the war ordered by their undemocratic regime."

Lori: "How long have there been war crimes trials?"

David: "The shocking thing is that they are an invention of the twentieth century, as far as I know. Liz, do you know otherwise?"

Elizabeth: "No, Nuremberg and Tokyo were the first, to the best of my knowledge."

David: "And that's 1945–1946, to try some mid-to-high level members of the Nazi Command and the Japanese imperialists. They were charged with 'crimes against peace,' that is, for the crime of committing aggression."

Nick: "I've always had problems with Nuremberg and Tokyo. Winners putting losers on trial; victor's justice, as they say."

David: "Well, who else was supposed to put them on trial? The UN was not yet set up, their own governments had fallen, and the Allies had control and responsibility over their territory. What was the alternative, not try them at all? Or shoot them on sight?"

Nick: "Have a neutral country try them."

David: "Easier said than done. Just think of the practical difficulties involved in the shattered postwar scenario. Plus, by 1945, who was neutral in World War II? Nuremberg always gets this bad rap, but people forget the context. Plus, they forget, or don't know, about the fair process of the Nuremberg trials. The accused were well-represented by good lawyers, the judges showed lots of integrity, and not everyone was found guilty. Pretty good for the world's first ever war crimes case."

Lori: "Isn't there a new court for war crimes?"

Elizabeth: "Yes, the International Criminal Court. Established in 1998 as part of the UN's International Court of Justice. Its goal is to avoid these accusations of 'victim's justice' by having a permanent court, judging all war crimes in all wars. And the judges wouldn't come from the countries involved in the case."

Lori: "Sounds great."

Elizabeth: "Yes, absolutely, but it's under-funded and already has a big backlog of cases. And America refused to support it, fearing politically motivated show trials of U.S. military personnel."

Nick: "Which is a legitimate concern."

Elizabeth: "I agree, as is the point that the United States already does arguably the best job of any country in internally prosecuting its own service personnel for alleged war crimes, or crimes against humanity activity."

David: "And you're part of that prosecution team."

Elizabeth: "Yes. And we really do stay objective and neutral, and we diligently investigate and prosecute. So, the ICC has nothing on us in that regard."

Nick: "I've heard my soldiers say they fear U.S. military lawyers more than the foreign enemy's military."

Lori: "But America must also object to the ICC on grounds of the American empire, and fears of the growth of world government, too, right? Not wanting a global government institution telling it what to do? Still retaining global hegemony?"

Nick: "Probably, but we've been down that road, conversation-wise."

Lori: "Right you are. OK, so, leaders of defeated aggressors should be deposed and put on trial, and now there's this new ICC to help with that. That sort of punishment I agree with completely; legal punishment targeted individually at those who deserve it. But why additionally punish the civilians?"

David: "Well, some people say not to do that. Others say it's a good idea to make the whole population feel the pain and sting and wrongness of war, even if they weren't to blame directly. The argument is that they may still be to blame indirectly, by having allowed such a government to come to power. Ultimately, the people of a country have to take responsibility for the quality and nature of its own government."

Nick: "Absolutely."

David: "Well, it sounds good in principle. I'm not sure it's so wonderful in practice. Like here, it may mean slapping sanctions on a whole country in the postwar moment, or making the whole country pay steep terms of compensation to a country it victimized by aggression. In either case, what results is a big decrease in economic resources for that country, precisely as it is trying to rebuild. Let me give you two historical cases showing this is not a good idea."

Nick: "Shoot."

David: "The classic modern instance of a postwar revenge policy was the Treaty of Versailles, which in 1919 sealed World War I."

Lori interrupted: "Wait. Should all wars end with formal, public peace treaties?"

David: "That tends to be the norm . . ."

Elizabeth: ". . . and the legal expectation, at least historically. And the publicity thing is vitally important; European countries, back under the old monarchies, would have secret 'reservations' or 'appendices' to public peace treaties, committing their country to private things which the public never knew about."

Nick: "Obviously impermissible in an open, democratic society."

David: "OK, getting back to the Treaty of Versailles. This vengeful peace treaty was quite punitive on Germany, which was blamed—and, I think, correctly—for beginning the war by invading Belgium in 1914. England and France, in particular, wanted to punish Germany for that, and they also wanted financial compensation for all their wartime costs. So they set up very steep reparations payments on Germany. They also took some German territory as booty, especially parts of the border area with France. Germany was extensively demilitarized, which is another element—the fourth?—in most conservative revenge policies. America, for its part, wanted, under President Woodrow Wilson, to see the growth and spread of democracy and so pressured Germany—formerly a military aristocracy, with a figurehead king or Kaiser—into experimenting with elections."

Nick: "And what happened?"

David: "Chaos. The key, really, were the reparations payments. They bankrupted Germany, ruining its economy. As a result, the people began to associate democracy with poverty, chaos, and failure. Then the Great Depression began, which only made the economic crunch more severe. Desperately, they looked to radical proposals to solve their problems."

Nick: "Hitler."

David: "Yes. It's commonly said that the Treaty of Versailles, by making things so difficult inside Germany, gave oxygen to an extremist like Hitler. He came along and said: 'I know the way out of this mess—ignore the Treaty. Let's stop making the reparations payments. Let's take back the territory the Treaty made us give up.' Simple, powerful 'solutions,' and some of the German people responded. And, as soon as the Nazis got a little bit of power, they used vicious tactics to take over completely and go from there."

Nick: "Starting World War II."

David: "Which is another reason why *jus post belum* is so important. It's my view that, *when wars are ended badly, they sow the seeds of future wars*."

Lori: "And you've just given us a nice illustration of that—and of how the revenge policy can create bad consequences and it is thus a bad idea. What about the other historical case you promised?"

David: "Well, now we get to Iraq, Lori, as you had requested. The case is Iraq 1, otherwise known as the Persian Gulf War. The war started, of course, very early in 1991, following Iraq's invasion and takeover of Kuwait in 1990. The goal of the war, on the part of the Allies, was to drive Iraq out of Kuwait and ensure Saddam wouldn't become a bigger threat to the region's vital oil supply."

Lori: "See! Oil!"

Nick: "Yes, Lori. We know your views about that. David is going to tell us of the peace treaty which ended that war."

David: "It, too, was a classic, conservative policy. Iraq had to cease, and declare stopped, all hostilities; it had to admit defeat; it had to admit its aggressive invasion of Kuwait was wrong, and apologize for that; it had to surrender any and all American POWs it had captured, in exchange for those Iraqi ones which the Americans had captured."

Nick: "I don't remember any war crimes trials, though."

David: "There were none. They left Saddam and his regime in power. That's one reason why it *wasn't* a postwar policy of reconstruction and rehabilitation. Furthermore, Iraq had to pay compensation to Kuwait, and to accept substantial demilitarization, so it would not aggress again."

Nick: "Makes sense."

David: "Yes, it does. As part of its demilitarization, Iraq had to agree to no-fly zones in the northern and southern parts."

Lori: "And why was that?"

David: "In the north, to protect Kurdish Iraqis from Saddam's central, Sunni-dominated government in Baghdad. And, in the south, to provide the same protection for Shi'ite Iraqis from Saddam. You've got these three groups in Iraq— Kurds, Sunnis, and Shi'ites—and they can't stand each other, historically."

Nick: "We've had some discussions about that."

David: "OK, just let me know when I'm repeating what's been said. The demilitarization limited the overall size of Iraq's military, forced Saddam to destroy many weapons he had, and then submit himself to a UN-run weapons inspection process which was targeted at identifying and eliminating all Iraqi weapons of mass destruction."

Lori: "And which found nothing."

David: "To the contrary, throughout the '90s the UN found tons of biological and chemical weapons, and destroyed them. Tens of thousands of tons. I'm not exaggerating. And the UN identified and dismantled many research networks which were devoted to the development of nuclear weapons."

Lori: "I'm confused."

Nick: "I'm not. Saddam *had* WMDs. The UN process confirmed that, and destroyed many. That's not the issue. The issue was whether the UN *found it all* when Saddam kicked them out, and refused to keep cooperating, in 1998. He said: 'You found it all; now get out.'"

Elizabeth: "Whereas the United Kingdom and the United States replied: 'You only kicked out the UN because they were coming close to finding the true hidden gems of your nuclear program.'"

David: "Precisely. And so there was a stand-off like that for a few years, but then 9/11 happened and changed things, making America more nervous about Saddam, and whether any WMDs were still there, and whether he had links to al-Qaeda and was about to give some WMDs to them."

Nick interrupted: "This we've spoken of and how controversial all that stuff was. But what I like about what you've said is drawing the clear connection between the terms of the peace of Iraq 1 and the outbreak of Iraq 2, twelve years later."

Lori: "Over demilitarization, and the weapons inspection process."

David: "That's not even the worst thing about the terms ending Iraq 1, though. The truly bad thing was the maintenance of the sanctions. After Iraq invaded Kuwait, America and its Allies slapped sweeping sanctions on Iraq to punish it. These sanctions prevented most trade between Iraq and the developed world, depriving Iraq of revenue. The goal was to reduce the resources at Iraq's disposal, so it would have less to create more trouble."

Nick: "Doesn't that make sense?"

David: "In theory, yes. In practice, less so. Especially when the sanctions are sweeping, as opposed to targeted. Targeted sanctions focus on punishing the elite decision-makers in a given society. Like freezing their foreign bank accounts, banning them from international travel . . ."

Elizabeth: "We talked about this, in connection with Clinton's targeted sanctions against the Haitian coup leaders in 1994."

David: "A fitting example of targeted sanctions. Sweeping sanctions, by contrast, harm all or most of the entire population in that country. They do this by shrinking that country's economy, by depriving it of specially needed goods and supplies, and so on. It's almost like economic siege warfare; the city, so to speak, has been surrounded, and you're squeezing them until they give."

Lori: "And Iraq was certainly squeezed in the 1990s."

David: "It certainly was and it backfired, as public opinion throughout the Arab world turned against America, as the evidence piled up as to how badly Iraqi civilians were doing. There was a ton of hardship borne by those, in many ways, least capable of protecting themselves."

Nick: "Whereas Saddam always found a way to take care of himself and his group."

David: "In fact, there's evidence that, far from weakening him, the sanctions strengthened his hold on power, as the people became even more dependent on political favors for income and access to resources with which to make ends meet. The sanctions policy was a failure on every level, in my view, and think of all the civilian suffering it created."

Elizabeth: "In violation of the rule of discrimination. And they had to have a second war, and give up on the sanctions policy, to get rid of him anyway. In retrospect, in my view, they should have gotten rid of him, and engaged in reconstruction and rehabilitation from the moment the first war ended in 1991."

Lori: "OK, now switch gears and tell us more about the rehabilitation policy."

Nick: "And why you're so convinced it's better than the conservative policy. Myself, I acknowledge the conservative policy isn't perfect—myself, I don't support sweeping sanctions—but, look, they were trying to contain Saddam. Leave him in power, but box him in. It's so much riskier to try overthrowing the regime and to create, by force, a Western-style society in an utterly different culture."

David: "OK, first thing to note is how much the rehabilitation policy still shares in common with the revenge policy. They both believe in: public peace treaties; some demilitarization; the exchange of POWs; and forcing the aggressor to apologize and to give up unjust gains. They both support war crimes trials for all those who deserve it. Where they differ is on: sanctions; reparations and compensation payments; and regime change. The revenge policy insists on reparations and compensation payments for reasons of punishment (which we've already discussed), whereas the rehabilitators just want to forget about trying to make them pay."

Elizabeth: "But why? It just seems so logically basic a requirement; if you started an unjust war, you must as a matter of principle and deterrence be punished for it."

David: "The rehabilitators agree, but they think the only permissible form of punishment is war crimes trials for the elite decision-makers. *That's it.* You cannot go on to punish the whole society by making them all pay for their regime's aggression. It violates the rule of discrimination, plus history shows it doesn't work; you only end up ruining that society and creating a huge number of new enemies. To beggar thy neighbor is to pick future fights."

Elizabeth: "OK, OK, you're persuading me; there's still punishment for the aggression but it's limited to criminal trials for those most responsible."

Nick: "But what about the society which the aggressor invaded, and may have destroyed? It has to reconstruct, too."

David: "The resources for its reconstruction simply have to come from elsewhere, either internally or perhaps internationally, through some kind of global compensation fund. Think, for instance, of America's response to the 9/11 attacks. Did we try to get compensation from the Taliban regime in Afghanistan—who supported and provided safe haven to al-Qaeda? No, we rebuilt the financial district in New York with our own money. A similar approach, in my view, should be done elsewhere."

Nick: "OK, now I think you've got me on that. And you already had me on the sweeping sanctions, and how they are cruel and stupid. But what you will not get me on is overthrowing the regime and reconstructing the whole society, Western-style. For two reasons: 1) you just cannot impose Western values successfully on a non-Western society; and 2) the costs and burdens on us are just way too big and demanding. Look at Afghanistan and Iraq, right now, as illustrations of both points; reconstruction is not working, and it's costing us too much, in time, treasure, and the lives of our young people."

Lori: "Wait, Nick. What's going on? I thought you favored the action in Iraq. You told me that Gil's not going to die in vain for a bad cause."

Nick: "First, Gil is not going to die. Dr. Leung and his team will save him. Second, I am always an American patriot. Third, I support the Iraq 2 war because I think the president was forced into it after 9/11 for reasons of minimizing terrorist risk, and because the sanctions weren't working against Saddam. But that doesn't mean I favor trying to turn Afghanistan and Iraq into little clones of Western, or American, civilizations. They are too different for that to work. I wish we would've just overthrown the Taliban, and Saddam, and said: 'Don't replace them with anything like them, or we'll be back,' and then left them to their own devices regarding reconstruction, in light of their own local values. Just as—as the good professor reminded us—we had to reconstruct Manhattan on our own after 9/11. At heart, I'm an old-style U.S. patriot who thinks George Washington was right when he told us to mind our own business, and stay out of foreign affairs unless we're attacked and absolutely must defend ourselves and protect American values and lives."

Elizabeth: "I think many Americans think as you do."

David: "I'm not one of them, though. That attitude may have made a lot of sense back in the 1780s, when countries were still developing and were still very separate. But history has changed that, and the way the world has developed—with all the interconnections and globalization—means, in my view, we can't indulge in nostalgia for American isolationism anymore. Plus, it's just callous to shrug your shoulders and say 'Now, you're on your own.' And I have to say, Nick, you're wrong when you say Western values can't be imposed successfully onto a non-Western society. The American reconstruction of Japan, from 1945–1955, proves otherwise."

Lori: "Tell us more about that case."

David: "Well, the Japan of 1945 was totally non-Western: socially, religiously, politically, racially, and ethnically, and so on. There was no prior experience with democracy (unlike in Germany). Yet the Americans, under Douglas MacArthur's leadership, created—over a period of ten years—a peaceful, stable, prosperous, successful, moderate democracy. Japan today is one of the best countries to live in, and is a very peaceful and productive member of the international community. America built a Western-style political structure. If it can be done there, it might be possible to do it elsewhere."

Elizabeth: "Like Afghanistan and Iraq?"

David: "In my view, no. But not because it can't be done in general, which is what one hears much too often. Japan shows that it *can* be done—the top-down, postwar creation and imposition of a Western social structure upon a non-Western society. And the fact that it *was* done in Japan, in my mind, shows that things like democracy, human rights, and prosperity are not just Western, but rather, are truly *universal* values. I mean, show me a person who doesn't want to live in peace and prosperity, who doesn't want security and freedom, and who doesn't want at least some say in how they are governed. The reason why the process which worked successfully in Japan—and postwar Germany, too—probably won't work in the Middle East is that there's a different context in Afghanistan and Iraq."

Elizabeth: "Say more."

David: "Well, before I do, let's say more about this process, or 'recipe' as I call it, and then we'll apply this recipe to Iraq and Afghanistan, and see how well things are going. Then I'll conclude by comparing and contrasting the present cases to successful past ones, such as Japan and Germany."

Nick: "Go ahead."

David: "Well, we've first got to get clear on the goal of postwar reconstruction. If you're going to impose deep social and political transformation upon a society in the aftermath of war, you need a clear vision of what you're trying to achieve."

Lori: "I myself think this has been missing in Iraq and Afghanistan."

Elizabeth: "Me, too, unfortunately."

David: "Myself, I think it hasn't been missing, I just think it's proving incredibly difficult to achieve. Mind you, I think mistakes were made elsewhere, as I'll say. The goal, as I see it, ought to be the creation *not* of a full-blown Western society, but rather, of what I call *a minimally just regime*. Such a political structure satisfies three vital, basic conditions: 1) *it is peaceful and nonaggressive* to other societies and indeed, to its own civilian population; 2) *it is legitimate* in the eyes of its own people, and the international community; and 3) it does what it can to *satisfy the human rights of its own people*. I think this has been more or less the goal with Afghanistan and Iraq, and America and its Allies have sought to apply historical best practices—developed from cases like Japan and Germany—to try to secure this goal and create these kinds of social conditions."

Elizabeth: "So what are the historical best practices, or this process or 'recipe,' as you call it?"

David smiled and said: "It's a ten-step recipe aimed at creating a society which is minimally just in the sense I just defined. I'll run through the steps quickly, and comment on how well each step has been received or accomplished in Afghanistan and Iraq. First: the winning, occupying force—the 'Reconstructor,' let's call it—must adhere to the laws of war and the rules of *jus in bello* during the process of taking down the regime in the 'Target' society. I think this was, on the whole, done well by the international Allies during the take down of the Taliban regime in Afghanistan. But I don't think the same can be said of Iraq."

Elizabeth: "Because of Abu-Ghraib, right?"

David: "Exactly. My impression is that, because Saddam was so widely loathed and feared within Iraq, many ordinary Iraqis—probably even the majority— were willing at least initially to give the Americans a chance at leading, or at least guiding, postwar reconstruction. But, as soon as those few bad apples in the U.S. Army did those things to Iraqi detainees, and it was splashed all over the media, it cost the Americans enormous credibility. A few bad apples spoiled the bushel, and suddenly the Iraqis became suspicious of the United States' intentions and how the occupier would behave. All that good will, gone up in smoke within a matter of days, and just because of the ill-considered, abusive actions of a handful of people."

Lori: "The next step?"

David: "Purge a lot, but not necessarily all, of the old regime; and prosecute those people who deserve it with war crimes trials."

Elizabeth: "But there've been no trials. Except for Saddam. And I don't think any have been held in Afghanistan for ex-Talibanis."

David: "And that may be a mistake or failure. But the Allies did purge the Taliban, though it was not eliminated and, in fact, there have been moves, or at least discussions, to invite the Taliban back into government in some Afghani regions."

Nick: "What? Outrageous!"

David: "Well, maybe. But just at the regional level in areas where it seems a cruel fact that they aren't going away and must be accommodated anyway. They won't be part of the national government. I'll talk about this more towards the end when I'll mention the Northern Ireland analogy."

Lori: "In Iraq, the United States purged and declared illegal the ruling Ba'ath Party."

David: "Yes, very effectively. Some even say too much, the United States sacked much of the Iraq civil service, since nearly all of them were Ba'ath Party members. But critics say the United States lost a lot of bureaucratic expertise when it did this—lost a lot of local knowledge, plus created bad will. Supporters of the United States policy said they were just following the recipe from Germany: 'De-Nazification.' But the critics reply that the Ba'ath Party was less like the Nazis and more like the Communist party in Russia, where plenty of bureaucrats joined the party purely for career-related reasons, and not for reasons of ideology."

Nick: "Interesting."

David: "Steps three and four are intertwined. Disarm and demilitarize the Target society—so as to make it more peaceful and less aggressive—but, if you do, you yourself as the Reconstructor must step in to provide security for the people in the meantime."

Nick: "Why? Why must we provide them security?"

Lori: "Yeah, why must my Gil go over there and help provide protection for them?"

David: "Several reasons. First, we were the ones who kicked over their regime. We're responsible. Second, as the winner, we have capabilities other, local players and agents, simply don't have. Third, it's a rough-and-tumble world, and some parts of it are even worse than others. If we don't provide security, then the society becomes vulnerable to invasion from nasty neighbors . . ."

Nick: "Like Iran, in the case of Iraq."

David: "Possibly, or the society becomes vulnerable to criminal elements within itself. What was Colin Powell fond of saying? It's like the Pottery Barn rule: 'if you break it, you buy it.'"

Elizabeth: "At least until it's reconstructed."

David: "Right. Now, this issue of security is massively important, as the evidence suggests strongly that security is absolutely necessary for progress on other postwar fronts, notably restoring economic growth. Think about it: if you're a shop owner; are you going to open up your business if there's a good chance you might get shot, or it might get blown up?"

Lori: "Of course not."

David: "So peace and stability must be restored. But have they been in Afghanistan and Iraq?"

Lori: "Nope."

David: "Well, it's complex. Let's start with Afghanistan. The capital area, Kabul, has been rendered quite secure by the international forces. So, some success. But the same cannot be said of rural Afghanistan, especially in the border area by Pakistan. Violence erupts regularly there; the Taliban is still a force there; most people think that this is where al-Qaeda and Osama Bin Laden are hiding; and often there is a multi-party exchange of fire, and military operations, involving the Allies, the Taliban, al-Qaeda, and the Pakistan Army. The border region is a total mess."

Elizabeth: "So, it's a mixed record on security in Afghanistan. Same thing with Iraq?"

David: "Yes, but I'm inclined to think it's a bit worse. One of the most controversial moves in the postwar era was U.S. Secretary of Defense Donald Rumsfeld's decision to disband the Iraqi Army in June, 2003, just after the regime fell."

Nick: "But that's an application of the disarmament principle, no?"

David: "Yes. That's what he was thinking. He was also thinking that American troops could secure the postwar situation better than they did. And that's not at all a comment on the *quality* of the U.S. soldier . . ."

Nick: "It better not be, Professor."

David: ". . . it's a comment on the *quantity* of U.S. soldiers over there. Rumsfeld subscribed to a view that *more could be done with less*, especially given America's huge advantage in military technology, weaponry, and cash."

Lori: "And he was wrong about that."

David: "It seems so. The evidence suggests that old-fashioned 'police presence,' as it were, is vital to securing postwar peace and tranquility in an occupied society. The number of 'boots on the ground,' as the Pentagon says, is vital. So, people think Rumsfeld made a huge mistake disbanding the Iraqi Army, as those soldiers could've been used to help keep the peace alongside the Americans. Of course, the top Iraqi generals and officers would've had to be purged, as they were too close to Saddam. But the rest, so the criticism goes, should've been folded into a new Iraqi security force under U.S. leadership. As it was, Rumsfeld sent them home and created a lot more bad feelings towards the Americans. Even worse, he let them keep their weapons, and many of them wound up in the Iraq insurgency, resisting U.S. occupation."

Nick: "Poorly planned. Blanket thinking, lumping all the soldiers together."

David: "And so, ever since, the Americans have been desperately playing catch up with security. Trying to provide it themselves, hiring some private mercenary forces to help out, and trying to train new Iraqi military and police personnel to take over. But the private security companies are even more distrusted by the people than the U.S. military, and the new Iraqi soldiers and cops suffer from three things: incompetence; corruption; and being targets themselves. They just cannot get it together over there, security-wise."

Nick: "But then the U.S. troop surge in '07 helped out."

David: "It did, admittedly. The U.S. soldier is still the one best positioned to provide security in Iraq. Increasingly, it looks like the insurgency might be diminishing in strength. But, even so, there are still two big security challenges over there: the first are the threats, and the reality, of neighbors interfering and intervening, and even supplying weapons and threatening to invade."

Nick: "Like Iran and Syria."

Elizabeth: "Don't forget Turkey—over the Kurds in the north."

Lori: "Yeah, what's up with that?"

Elizabeth: "The Kurds are the majority in northern Iraq. But there are also Kurdish people in Eastern Turkey. They all want to secede from their current countries and form an independent Kurdistan. In Turkey, these Kurdish rebels have a separatist force which uses violence, targeting Turkish soldiers. So, Turkey is very hostile to the Kurds and might invade Northern Iraq as part of its campaign to keep its own Kurdish separatists under control."

David: "The second security challenge comes from the other big internal rivalry in Iraq: Sunni-on-Shi'ite violence. Intercommunal warfare. It looked like such rivalries and attacks would create a civil war, and people openly talked of that in 2006 and 2007. But the U.S. troop surge seems to have helped calm things down a bit."

Nick: "Yeah, but for how long? U.S. troops will start coming back home in very late 2008. So they say, anyway."

David: "And that's just it. The U.S. military is, in my view, trapped there; if it pulls out, chaos may erupt. Utter chaos: internal civil war; international invasions from neighbors; and so on. Yet, while there, the United States doesn't seem that capable of keeping the peace, as there is still a shocking amount of violence. Can't pull out, yet can't perform well while staying in."

Nick: "Damned if it does; damned if it doesn't."

David: "It's a very difficult situation. I don't know the solution."

Lori: "What about increasing the size of the U.S. military presence there? Your whole 'boots on the ground' thing."

David: "There's compelling evidence that an occupier can, over the short- to medium-term, impose peace with raw quantity of military presence (just as they've found police presence dramatically cuts crime in our own society). Now, over the long-term, obviously, raw force is not enough; the people must endorse and support the social structure. The problem with that is, I think, there's no political support for the kind of dramatic surge that would need to be in place to impose peace in Iraq."

Elizabeth: "People are tired of the war and its problems."

Nick: "War fatigue."

Lori: "And people like me are also sick of losing loved ones, or hearing about it. Yet, if it were to work, I might feel better if America was, finally, genuinely able to improve Iraq. I'd feel a lot better in the end about Gil's sacrifices if they were on the side of a successful resolution."

Nick: "I guess only time will tell. We'll see what Congress, the president and the U.S. voter have to say and decide about it."

David: "Moving forward with the 10 step recipe to step number 5. And here, unlike security, is where we can see some clear American-led progress. It's this: the occupier should work with a representative cross-section of locals on developing a new constitution, and political structure, which fairly represents everyone's interests."

Nick: "And we have done that: new constitutions, plus free and fair public elections, in both Afghanistan and Iraq."

Elizabeth: "In countries where there were no such things before."

David: "A real—and even amazing—achievement, absolutely. Americans are good at democracy, new constitutions, and free elections."

Lori: "But we could use some help on securing peace and stability, and cutting down on violence. Oh, and curbing our greed for the oil."

Nick: "As he admitted, Lori. Let him move on to these further points, OK?"

David: "In spite of the progress, there are still some political challenges, though, in both nations. In Afghanistan, Hamid Karzai has been a stable leader and presided during this remarkably difficult time. But I've heard people joke that he is more like the mayor of Kabul than the president of Afghanistan, referring to this significant split between the capital and the rural areas. Plus, there's the whole issue of the warlords."

Lori: "Yes, who are they? You always hear of these people."

David: "Afghanistan is not an advanced society. It is still very much based on clan, or extended family, structures. At the head of the clan is a middle-aged man. His job is to protect a family's traditional turf, and he does so with his own family-composed little army or police unit: his sons, brothers, grandsons, nephews, and so forth. These guys are all still armed—they could easily turn into security threats. For now, they've been content to support Karzai."

Nick: "In exchange for what?"

David: "Doing what he can to prevent the Americans from dismantling their traditional social structures and, above all, keeping his hands off their poppy fields and drug operations."

Lori: "What?"

David: "Afghanistan is one of the world's poorest countries. Very few things grow there. But poppies grow really well, and poppies are the main ingredient in opium and heroin. Many of these warlords are essentially drug lords, providing drugs around the world."

Elizabeth: "I heard the one good thing the Taliban tried to do, back when it was in power in Afghanistan, was to attack the drug trade."

David: "But now they are gone, and the drug trade is back, and in fact it is an armed drug trade, and what you have is this very sad, potentially explosive, situation of several hundred territory-based armed drug dealers saying to Karzai, the president: 'mess with us, and we'll withdraw our support, and society will go back up in smoke.'"

Nick: "Cripes. I had no idea it was so precarious."

David: "Even worse is how the Taliban are making an effective come-back in some rural regions, to the point where Karzai is seriously considering offering them regional power-sharing deals."

Nick: "Isn't that like dancing with the devil? Didn't Hitler teach us that you can't compromise with fascists? Give them an inch, and they'll take a mile."

David: "Maybe, but Karzai's in a very tough spot—as I've just described—and so I think he just feels it's a practical political concession that essentially just recognizes what he can't change anyway; in some pockets of the country, the Taliban are in control. Plus, I also think he may be thinking of Northern Ireland."

Elizabeth: "How so? My mother's family is Irish."

David: "Well, there's evidence that the fighting between the different communities in Ireland stopped when the extreme players, who used violence and terrorism, were brought into the mainstream political process. When they *politicized* their cause, they were forced to *moderate* their views, as generally the majority in any given population is not extreme but moderate."

Nick: "But, c'mon, the Taliban—these violent Islamic extremists—are going to moderate their views once they are forced to share power? Isn't that wishful thinking on Karzai's part? Again, I come back to Hitler: when he got a small taste of, and access to, power—after his first few political wins—he *didn't* become a responsible, moderate political statesman. He just wanted, and got, more and more and more. Karzai's playing with fire."

David: "Maybe. Let's now consider Iraq's political situation. Here, too, there's been clear progress, with new constitutions and free elections. Plus, did I mention women have been elected both in Afghanistan and Iraq for the first time ever in history? But the big lingering problem has to do with the relations between the three groups: Kurd, Sunni, and Shi'ite. The Kurds don't even believe in a unified Iraq. They are just waiting for the opportunity to secede and join with the rest of the Kurdish people in an independent Kurdistan. But, for now, they *do* support the existing constitution in Iraq, as it gives them much more regional autonomy—much more effective control over their own affairs—than they've ever had before."

Elizabeth: "So, the new constitution over there is federal?"

David: "That's right. Just like in America; the various regions, or provinces, or states, retain many political or economic powers, and the central federal government only has a small amount."

Nick: "The only way to keep them together is, basically, to keep them apart."

David: "Well said, Nick, and that's exactly true and proper, given their historical antagonisms. But the real threat to political stability and progress in Iraq isn't the Kurds: its Sunni-Shi'ite relations."

Lori: "And we know a bit about this; the Sunnis, centered around Baghdad, are the minority, yet they are used to having power in Iraq. This has made the Kurds and Shi'ites feel aggrieved and resentful. The Shi'ites, for their part, are the majority, yet feel historically under-represented."

David: "Exactly. So, the Shi'ites actually love the new democracy, as they are the majority and thus can finally count on having the power."

Nick: "Whereas the Sunnis hate everything which has happened since 2003, as they've lost their control."

David: "Bingo. This is one of the two huge political issues for Iraq right now; keeping the Sunnis committed to, and involved in, power sharing in the new Iraq. Much of the insurgency, I'm told, is made of disaffected Sunnis. The Sunnis are almost twenty-five percent of Iraq's population, and thus they can't be ignored, even if there is some justifiable historical resentment against them. Plus, they tend to be a bit more moderate in their religious views, which fits in more with the kind of modern society America is trying to help develop."

Elizabeth: "What's the other big political issue?"

David: "It's both political and economic—the fair division of oil revenue between the three groups. Afghanistan's huge economic problem, we've seen, is the lack of natural resources and the development of a violent drug economy, dominated by warlords. How to grow a decent, legitimate, peaceful, non-narcotic economy? Iraq at least has the oil, plus a better educated population. But the new Iraqi constitution stipulates that oil revenues must be fairly distributed, and I quote 'with an eye towards redressing historical injustices,' which everyone knows clearly commands *less* for the Sunnis and *more* for the Kurds and Shi'ites. The Sunnis are thus furious, both politically and economically."

Nick: "But don't they deserve it?"

David: "Well, maybe. But, even if they do, how will that help Iraq move forward? If one large segment of society is left out, or feels bitterly aggrieved, this can only create huge problems."

Nick: "Or they could just learn to suck it up, and deal with their hurt feelings."

Lori: "Or, listen, why don't they just carve Iraq up into three new countries, if there are these deep, bad feelings?"

David: "Several reasons. One: it could seriously escalate into a broader regional war involving neighbors, especially Turkey, Syria, and Iran, but also possibly Saudi Arabia. Two, that actually wouldn't solve the 'division-of-oil-revenue' issue. Think of a divorce; it solves some things, like if you can't stand living together any more. But some issues remain and linger, like alimony payments and child support payments. It would be the same thing for Iraq: the groups would say, 'well, we all used to share this oil asset in common, but now we have to figure out how to split and share it,' and then you're back to where we are right now. Partition might solve some political issues—if they could figure out how to avoid war—but it will solve none of the economic problems. Thus, obviously not a winning strategy."

Nick: "So what's the solution to the political and economic problems in Iraq?"

David: "I don't pretend to know."

Lori: "The next steps?"

David: "Well, step number 6 is to forego the reparations and compensation payments approach of the revenge model, and instead to invest in, and rebuild, the defeated society. This worked brilliantly well in postwar Japan and Germany, through the Marshall Plan."

Nick: "And which we've done yet again today, both in Afghanistan and Iraq."

David: "To the tune of billions. And it's great; look at how hard reconstruction is right now, and then imagine how utterly impossible it would be if, while reconstructing, we were also sucking money out of these countries through sanctions and/or reparations requirements. *For the sake of the future, we must learn to let go of the past.*"

Elizabeth: "At least when it comes to whole societies. When it comes to war crimes, committed by particular individuals, those we should *never* let go of, until they have been punished."

David: "Agreed. So, we've done this investing, but then, as I've mentioned, it hasn't prevented these societies from still having huge economic problems: in Afghanistan, it's the drugs and general poverty and underdevelopment; in Iraq, it's the division of oil revenues, plus the security chaos hampering economic growth. Both countries still need substantial rebuilding and infrastructure repair from the various wars they've been in since 1979—and they both suffer from massive unemployment which probably creates violence, as too many angry young men with nothing productive to do often . . . causes trouble."

Nick: "Solution?"

David: "More cash and international aid; the application of the latest tools and know-how regarding economic development. Myself, I like Amartya Sen's book, *Development as Freedom*, where he talks about nurturing capabilities in people, capabilities which expand their choices and freedom in life. Much development research stresses the need to invest in the education and economic opportunities for girls and women in particular."

Lori: "Ha! The feminists were right; give the money and the power to the women, and there's your solution to all this war crap."

David: "I fear that's an overstatement, but much development research does show huge benefits from investing in women. A lot of international aid in

Afghanistan, for example, goes into building schools for girls, and providing seed money to small-scale women entrepreneurs who wish to start their own small businesses, like carpet weaving or a bakery shop, and so on."

Elizabeth: "That all takes time to see good effects, though, right? A long time to get through school, or to ramp up a successful business."

David: "True, and in a way, postwar reconstruction is a race against the clock. James Dobbins has come out with two terrific books on postwar reconstruction and nation-building: the first on America's involvement in such; the second on the UN's involvement."

Elizabeth: "Is there evidence one has been better than the other at leading and generating postwar reconstruction?"

David: "No. Both have had wins and losses but, in my view, the UN has never had a success like postwar Germany and Japan. So, there's no evidence that the more players involved with postwar reconstruction, the better it goes."

Nick: "That's because too many cooks spoil the broth."

David: "Maybe. In any event, Dobbins suggests the key is the relationship between the war winner and the local populations, and the enduring commitment *of both* to successful reconstruction. For its part, the winner can't walk away, and wash its hands, too soon. For theirs, the local population has to overcome its own internal divisions and want new institutions which are better and which work."

Lori: "But you were speaking of a timeline."

David: "That I was. Dobbins suggests there's a seven to ten year postwar window of opportunity and, if reconstruction can't be successfully rooted and well on its way by then, it will fail and either another war will break out or the target society will fall apart as a 'failed state.'"

Lori: "Seven to ten years from when? When does the clock start ticking?"

David: "From the collapse of the old regime. So, in Afghanistan, the Taliban fell in January 2002, so we're looking at good success by 2012. In Iraq, Saddam fell in May 2003, so we're looking at 2013. Widespread success in terms of the ten elements in this recipe and in terms of achieving the kind of minimally just society I mentioned before: peaceful and nonaggressive; secure, with a legitimate political structure; and sincere efforts at realizing the human rights of the people."

Nick: "So, we still have some time and, as you've admitted, there have been successes."

David: "Yes to both, yet the remaining challenges shouldn't be underestimated. Dobbins says the average citizen in the defeated country must, within ten years post-regime collapse, feel their life is concretely better than it was before the war began. And, by concretely, Dobbins stresses two things: *feeling more phys-ically secure* from violence, chaos, war, and tyranny; and *feeling more econom-ically prosperous.* Otherwise, they will turn their backs on reconstruction, and search for other options."

Lori: "Like war, once more. Next step?"

David: "Step number 7 ties in with this last one. The benefits of the new, re-constructed order must be both concrete and widely distributed, and enjoyed, throughout the population. The benefits of the new social order cannot only be philosophical abstracts like democracy and freedom; they must also be concrete, in particular in terms of physical security and economic growth."

Lori: "We're not there yet, in either country. So it's still very dodgy."

David: "I agree. Also, there's the issue of the *distribution* of benefits. There can-not be an elite privileged insider group seen to get most of the benefits, else they will be branded as traitors and agents of occupation. We've seen there's the whole related issue of the Sunnis in Iraq."

Lori: "Plus there's the issue of the American oil companies, and private security companies, making out like bandits from reconstruction. The Iraqis must hate that."

David: "Step number 8 is to allow for 'civil society' associations to flourish. These refer to any and all non-state or nongovernmental associations: from busi-nesses to churches, from sports leagues to the free press, and so on. Here, there have been big gains both in Iraq and Afghanistan, which were both dictatorships under the previous regimes."

Nick: "See Lori, Gil's been helping out with concrete gains in human freedom around the world."

Lori: "Yes, and look at what cost to himself . . . anyway, let's finish off steps 9 and 10."

David: "In Germany and Japan, it was necessary to revamp radically the educa-tional curricula of schools, as the students were taught these horribly twisted doctrines of racial supremacy, fascism, and so on. So, educational reform is step number 9; set the stage for enduring social change by more properly educating the next generation."

Elizabeth: "Sounds good. Have we been doing that in Iraq and Afghanistan?"

David: "Not at all in Iraq."

Nick: "Big mistake."

David: "Probably, but look, there's problems enough with security, politics, and the economy. We have been trying it, in a limited way, in Afghanistan."

Lori: "By educating girls and women, like you said."

David: "Exactly."

Elizabeth: "I think that may actually be the postwar measure I most support."

David: "The final step is for the winner/reconstructor to exit at the right moment. It's about exit strategy; don't leave too soon, or the new society will fall apart and reconstruction will fail; yet don't leave too late, because then you're creating dependency and, at the extreme, you might essentially be engaging in imperial conquest and making the target society, effectively, a colony of your own society."

Nick: "And, obviously, with these two current wars we don't yet know if we'll get the exit strategy timing right, as we're not there yet."

David: "And Dobbins suggests we shouldn't be getting out, in either case, until about 2012–2013. So, I must admit, I get really nervous when I hear people talk about massive U.S. and Allied withdrawals of troops before then."

Lori: "But, they're sick of U.S. casualties, and they think reconstruction has failed. So why not get out? Plus, just think of the dollar cost of these wars and the other, better things we could've done with all those billions: health care; better schools; repaired crumbling infrastructure . . ."

David: "But if we're in for a penny, we're in for a pound. We *started* postwar reconstruction in these societies, and we've got to *see it through* successfully. It's our responsibility. And not just that: if we do see it through, we will improve the national security of the United States."

Nick: "Damn right."

Lori: "How so?"

David: "I believe in what is called 'the democratic peace thesis.' The Enlightenment-era philosopher Immanuel Kant was the first to suggest it, and Michael Doyle has recently got a lot of attention for it. It goes like this: good societies don't go to war against each other; or, more precisely, rights-respecting democracies don't go to war against each other. So it follows that creating more rights-respecting democracies will increase international peace, and thus American security."

Lori: "But isn't that too optimistic? When I think of Iraq, I think of Vietnam; America couldn't get out yet nor could it win. It is just an impossible trap, in a brutal part of the world, where we simply cannot create the kind of just society you're talking about. And the greed for oil just taints everything."

David: "Whereas that attitude seems too pessimistic to me. Don't get me wrong; I'm aware of the challenges, and I'm not sure success can be achieved in the required time frame. But the fact is that some successes have been had, and there is still substantial time left. I think it's an issue of resolve and resources and, if those come together, perhaps we can pull the rabbit out of the hat. Surely, if Nazi

Germany and Imperial Japan can be reconstructed, then so too can Afghanistan and Iraq."

Elizabeth: "But you yourself said there are important differences between those cases."

David: "And there are. I haven't tried to hide or downplay the problems. For me, I guess it comes down to an issue of when to diagnose failure, which would justify leaving for good. I don't think we should do that until the ten year time frame Dobbins suggests. I'm for commitment, presence, and creative solutions, until then. We give it our truly best shot."

Nick: "And I'm with you."

Lori: "Whereas I think it's over, and we should all go home."

Elizabeth: "Whereas, for my part, I must admit I'm not sure."

Lori: "Another thing."

David: "Please."

Lori: "This whole 'minimally just,' or 'rights-respecting democracy,' thing. How is this a justified goal or standard to aim for, with postwar reconstruction? Isn't this just the imposition of Western or American values?"

David: "People say that but I don't believe it, or think it withstands scrutiny. I think the requirements I've laid out for a minimally just society are not just Western, but rather, are truly universal. They are desirable in the eyes of every reasonable person. Who doesn't want peace and nonaggression? Who doesn't want to live under a legitimate social structure which protects everyone's fundamental interests? Who doesn't want their human rights satisfied?"

Elizabeth: "How do you understand 'human rights'?"

David: "They are the most important political values, in that they are the most important claims any of us make on each other and on the social institutions we share. Human rights are basic entitlements we all have to the things we need to live at least minimally good lives. We should all, as a baseline proposition of justice, have the opportunity to live lives we ourselves see as being minimally good and thus worthwhile. Otherwise, what's the point of being alive? And, to enable a minimally good life, I say we need five things: personal security; material subsistence; personal freedom; elemental equality (or nondiscrimination); and social recognition as a person and rights-holder."

Lori: "I've always thought of human rights as Western."

David: "Well, the idea first comes out of the West, but that hardly limits its application and relevance to Western civilization. Every rational person wants their human rights respected and realized, so they can live a minimally good life. Moreover, societies which respect human rights are better, happier, more stable, and prosperous than those which don't—so, we'll improve people's lives by pushing for human rights. Finally, we'll increase peace and decrease war by focusing on the fulfillment of human rights. That's why postwar reconstruction measures must include human rights; that's how we ensure the war, and postwar, investment and cost was truly worth it, by bringing into being a better world."

Nick: "Amen. You've got me back into the fold."

Lori: "Not me. Too optimistic."

Elizabeth: "I wish I could agree, but for now, I think I still side with the conservative postwar position; no reconstruction as it's costly and risky."

Just then, Dr. Leung entered. All heads turned to him. He wore a very serious expression and obviously had come straight from the operating theater. With him was a short Hispanic man in uniform.

Dr. Leung: "Lori, I'm very sorry. Gil's taken a turn for the worse. Everyone, this is Father Diego Cortez, an Army chaplain here on base. Lori, I gather from the hospital admission forms that Gil was Catholic. I'm deeply sorry to say I've called in Father Cortez here to administer to Gil the last rites."

Chapter Eight

Peace in this Life

Lori had a horrible night. She watched on, sobbing, while Father Cortez performed the Catholic ritual of the last rites on Gil. Gil was then wheeled back into the surgical theater for emergency surgery aimed at preventing the event which now seemed increasingly inevitable: his death from shrapnel wounds inflicted in Baghdad.

Lori and her uncle, Nick, had spoken on the phone with the rest of Gil's relatives, including his parents. They were too old and ill to make the flight over, but were fully informed of what was going on. They were devastated, and prepared for the worst.

Elizabeth anxiously wanted to continue on to Baghdad to get on with her investigation, though she felt—given what information she had—no charges or disciplinary procedures would be necessary. It was either an accident, or wounds inflicted by the enemy; very sad, in either event, but nothing legally relevant to pursue. Yet, in spite of her desire to do her job and get back home, she felt in limbo since the General was supposed to be her escort, and now he had this family crisis to attend to. She felt she had to wait for him to decide when they would leave. Plus, she didn't want to offend Lori, to whom she had grown attached. "I would want support, were I her," Elizabeth thought to herself, as she got up to have breakfast, "and so here I am until the General decides what to do."

When Elizabeth entered the mess hall for breakfast, she saw David and Father Cortez in earnest conversation. She grabbed some hot tea, a banana, and a bagel with cream cheese, and joined them.

Elizabeth: "What are you two discussing?"

David: "Father Cortez was telling me about his pacifism."

Elizabeth: "Pacifism? But, you're a U.S. Army chaplain. You work for the Army: how can you possibly be a pacifist?"

David: "My question exactly. Plus, how can you be a pacifist when you're Catholic? Doesn't the church believe in, and support, just war theory?"

Diego: "Well, my higher and true calling is to God. God first, then country. And I believe Jesus Christ calls me to bear witness, in thought and deed, to the hope for a more peaceful and loving world. Plus, the U.S. Army is filled with men and women who have deep spiritual needs—like Gil right now—and I can help serve those even if, in my heart of hearts, I believe that war is always wrong. As for my Catholicism, you technically shouldn't call me 'Father.' I used to be a Catholic priest, but am no longer, as I fell in love, got married, and had children. Priests cannot marry, so off I went. Plus, as David notes, my pacifism—and some of my other beliefs—put me a bit at odds with church teaching."

Elizabeth: "Wow. How long ago was that?"

Diego: "A long time ago, maybe seventeen, eighteen years now. I just think that, in my early '40s, I had an overwhelming desire to become a father. I met the right woman. I was conflicted in my conscience about some aspects of church teaching, and so left the priesthood."

David: "Takes a lot of guts; good for you."

Diego: "Thank you. I don't regret it; my wife and I have four kids, and are still happily married all these years later. I still serve Jesus Christ in my new role as Army chaplain."

Elizabeth: "Does the Army know about your pacifism?"

Diego: "Yes, but they respect it as a private issue of conscience and freedom of religion. I don't try to convert soldiers to my way of thinking, and so what do they care about my personal beliefs? I tell people if they ask me, but it's not like I'm out undermining the morale of the troops by preaching pacifism."

Elizabeth: "But, forgive me, why *aren't* you doing just that? If you truly believe in pacifism, why wouldn't you try to convert people to your point of view?"

Diego: "My role is to bear quiet witness to a faithful, fruitful, peaceful life. The best way to convert people to your way of life is to serve as a role model they wish to mimic."

David: "But why be in the Army at all? You could do that in any number of jobs—social worker, international aid worker—or become a Protestant minister or reverend, presiding over some pacifist congregation."

Diego: "Well, why do any of us wind up where we do? It's a complex combination of factors. I'm in the Army for two reasons: my wife was in it before I was; and both of us are children of illegal immigrants from Mexico. We love

America, and want to pay it back. So, I wound up playing a religious role within a secular institution of the U.S. federal government."

Elizabeth: "Wow. Life and its twisting, turning paths."

David: "I think it's great, Diego. You serve your country, yet still get to be true to your conscience. You've got a great wife and four kids. And you perform a valuable service to servicemen and women occupying stressful and important jobs. You still serve God. It's a nice balancing of many different interests, and sides of yourself."

Diego: "Thank you. I'm always just so saddened when called upon to administer last rites to soldiers like Gil."

Elizabeth: "Wait. Since you're not a priest, can you actually administer last rites?"

Diego: "Technically, no. But . . . uh . . . I don't perform the full sacrament. I won't get into the details. The soldiers and families don't mind, though. In fact, none of them have ever asked, probably since very few of them have seen the sacrament before, and they are just so saddened and fearful that they simply need some religious comforting."

Elizabeth: "I'm sorry to ask. I don't mean to be rude."

Diego: "Not at all. But performing last rites to dying soldiers always reaffirms my pacifism."

David: "You feeling up to chatting about that in greater detail? Do you have time?"

Diego: "I do. But I'm performing a service later in the afternoon."

David: "Well, we'll promise to let you go by then. Tell me why you're a pacifist."

Diego: "First and foremost, for religious reasons. I believe in God and his only son, Jesus Christ, born of the Virgin Mary. My main duty in life is to obey the commands of God. And, you must admit, I have been fruitful and multiplied."

They laughed.

Diego: "So, I obey God's commands, first and foremost. I believe in the commandment not to kill. I also believe that I should, to the extent I can as a flawed and sinful creature, mold myself after the life and teachings of Jesus Christ. Nowhere in the *New Testament* does Jesus endorse violence or warfare. In fact, he criticizes both. He preaches instead love and forgiveness. Love your enemies. Forgive those who trespass against you. Turn the other cheek. Those who live by the sword will die by the sword. Jesus was the Prince of Peace, and I must remain faithful to his example, and condemn all violence and warfare."

David: "Great, but can you give *nonreligious* reasons for *others* to believe in pacifism? Or do people have to be religious to believe in pacifism? Or, even more narrowly, a certain kind of Christian?"

Diego: "No, there are very good secular reasons, to which rational people of any—or no—faith could agree. I'm just saying why *I'm* a pacifist, since you asked. Religious reasons trump all, for me."

David: "Fair enough. I just think that, when we talk about *public* policy choices—such as whether or when to go to war—we *cannot* offer religious reasons. Those are *private* issues of personal faith and conscience, and we can't expect other people of different upbringing and faith to agree with us. We have to stick with what John Rawls called 'the public use of reason.'"

Diego: "I know, and agree. In public life, we must offer public reasons, which is to say, justifications for action and policy that can appeal to people of any and all—or no—religious faith."

Elizabeth: "Amen to that."

They laughed.

David: "So Diego, do you have any such public, secular reasons as to why I should be a pacifist, and not a just war theorist?"

Diego: "Well, let's lay the foundation. Pacifism and just war theory agree it is both possible and necessary to judge war from a moral perspective. In that they both agree, as opposed to the ethical skepticism of the realist."

Elizabeth: "Right."

Diego: "But then pacifists and just war theorists differ in that, when it comes to the moral judgment of war, just war theorists say war can sometimes be morally permissible whereas pacifists say war is never morally permissible."

Elizabeth: "Never?"

Diego: "Not ever. I'm always amazed when people tell me they are pacifists, yet at the same time say: 'But I do support the war on terror; and I did think beating the Nazis was a good thing.'"

David: "Don't forget beating the slave-owning Confederates. And the godless Communists."

Diego: "Yes, yes, you get the picture. But there's no such thing as selective pacifism. If you endorse some war for moral reasons, you are in fact a just war theorist. Pacifism—true commitment to peace—requires *categorical* opposition to *all* wars . . . to war *as such*."

Elizabeth: "OK, good. Pacifists always think war is wrong. But why? How can they, for example, deny that World War II, or the American Civil War, were morally good wars? Wars worth fighting. Wars of compelling cause, and which improved the world?"

Diego: "Here's how I think of it. Now, forgive me, but I do have a doctorate in theology, so this may sound like a lecture."

David: "That's OK: I've got a doctorate myself and Liz is a lawyer. We're not afraid of learning."

They laughed.

Diego: "Well, not so much learning as listening. I've got a list of three reasons in favor of pacifism, based on three perspectives on the essence of ethics, morals, and justice."

Elizabeth: "Go ahead."

Diego: "First, and my favorite: virtue ethics. Aristotle founded virtue ethics. Its core proposition is that you should live your life—that is, think and act—trying to develop the virtues to the fullest extent you can. The ideal is the pursuit and achievement of human excellence, to the best of your ability."

Elizabeth: "And what are the virtues?"

Diego: "The virtues are traits of character which benefit both self and society. Such as intelligence, or kindness, or courage, or a sense of humor. Practicality and moderation are two more. They are difficult to develop, and we all praise them in others. They require conscious effort to attain, and must be used or exercised in a habitual way, so as to become truly part of oneself."

David: "You are what you do on a habitual basis."

Diego: "Absolutely. That's your character. Anybody can come up with one right answer, or one funny joke—but it's the people who can do so consistently whom we call smart and funny. Same thing for moral character; it's about developing in oneself, and helping to develop in others, a pattern of habit which displays the human virtues or excellences. And here's the tie-in to pacifism. Pacifism is called for, from a virtue ethics point of view, because fighting, violence, and war are not human excellences. They are not virtues; rather, they are vices. These things do not form part of what we, and Aristotle, would call a good—much less best—human life. War is about killing and bloodshed, anger and violence, hatred and destruction. It causes terrible pain and agony, lasting decades or even centuries. It is in no way expressive of an excellent human life, nor can it help to create such. Thus, it is always wrong. We must always strive for peace, as that is expressive of the good life; solving problems without violence, anger, and destruction. In sum: peace is a virtue, and war a vice; so virtue ethics always endorses the former and rejects the later."

David: "Nicely put. Eloquent. I have some comments, questions, and challenges, but why don't you put the other two reasons on the table, and then we can debate?"

Diego: "Gladly. The second reason comes from a second major tradition of thought about morality: deontology, or duty-based ethics. Here, the core concept is not of virtue or excellence, but rather, of duty or obligation. The essence of ethics is for us to do our duty. For example, obey the Ten Commandments, of follow the Golden Rule: do unto others as you would have them do unto you. The connection to pacifism is this: war is always wrong because it violates some crucially important moral duty. Consider the application to the Golden Rule; you should never go to war because you would never want to be warred upon, right? Don't inflict on others what you would not want inflicted upon yourself."

David: "I asked for secular reasons, though, Diego, not the Golden Rule of Christianity."

Diego: "OK, I have other duties to offer you. But, before I do, let me just say that the Golden Rule is still an excellent moral rule regardless of whether you are Christian or not. First, it is easily applicable to any choice people or communities face. No matter what situation you're in, you can ask yourself: 'How would I, or we, like to be treated?' Second, it is so clear and understandable. For example, with my kids, they understand it—even my four-year-old granddaughter understands it. She cannot understand Aristotle's views on excellence, but she totally gets it when I ask: 'How would you feel, if I did that to you?' Third, it draws people out of themselves and gets them to empathize with others. An excellent moral quality regardless of religion. Finally, the world would be a better place if everyone—Christian or not—behaved with respect for the Golden Rule. I firmly believe it is all you really need to know and consult when wondering how to treat other people. Do you think there would be war if people *did* obey the Golden Rule? Wars get started precisely because people refuse—out of ignorance, egotism or hatred—to extend to others the same kind of treatment they themselves want. If everyone treated others the way they themselves wanted to be treated, there would be no wars."

David: "Again, eloquent. And some persuasiveness to it, I grant you. The Golden Rule probably forbids aggression and violent power seeking, and I think these motives cause most wars. But I still want some *secular* duties."

Diego: "OK, how about this: do not kill another human being. What could be more basic to a system of ethics than that? It would be total chaos if we didn't have a rule like that."

Elizabeth: "What about killing in self-defense against a murderous attacker?"

Diego: "Two wrongs don't make a right."

Elizabeth: "But, if I kill a criminal in legitimate self-defense, then that is *not*, in my view, a second wrong. I am entirely right and justified in using killing force, if need be, to defend myself. His original attack is the one and only wrong, my self-defense is a right."

Diego: "What about the criminal's right to life?"

Elizabeth: "He forfeits that right when he tries to kill me."

Diego: "But you can't forfeit your basic human rights. They are natural endowments, given to you by God as one of his children."

David: "Again with the religion, Diego."

Diego: "Sorry but, look, I'm a priest. Or was. Hard habit to break."

Elizabeth: "I don't believe human rights are natural properties, or things we're really born with. They are the product of the social contract—we agree to recognize these rights of each other, in exchange for their respect of our own. If someone doesn't respect our rights, we don't have to respect theirs. I mean, take a criminal. When he commits a crime, he takes himself out of the social contract. He's broken the deal. And so he forfeits his right to be a free man at large in society, and we're justified in throwing him into prison if the crime calls for it."

Diego: "I disagree. But that's a very foundational disagreement about the nature of human rights and where they come from: you say from social consensus whereas I say from God. OK, look, here's another secular duty for you: the duty not to kill *innocent* people. You can't pretend you have the right to kill innocent people, that is people who *aren't* criminals or murderous attackers. Indeed, killing an innocent person is itself murder. There's no ethical rule more clear or compelling than that."

David: "And the connection of that duty to the wrongness of war?"

Elizabeth: "I know what he's going to say . . ."

Diego: "Well, let me say it, then. War kills innocent people. Civilians. The sick. The elderly. Young children. People who have absolutely nothing to do with the war. People who can't remotely be blamed for the war, and have done nothing to deserve death."

Elizabeth: "We had a previous discussion about how, in modern warfare, civilian casualties outnumber military casualties."

Diego: "And there's the true horror of war for you; it results in the death of innocent people. Nothing which does that can possibly be considered morally right or justified."

David: "OK, I've got a lot to say about that. I agree that that's a very important argument. But let's get the third antiwar reason on the table."

Diego: "It's this: consequentialism. The more recent ethical tradition, according to which actions and policies are to be judged according to the consequences (or results) they generate. In particular, I think of the main kind of consequentialism, utilitarianism. The utilitarians—like John Stuart Mill or Jeremy Bentham— said that the main thing each of us should do with our lives is to do what we can to make the world a better place. And, by better, they meant happier. Through your actions, make the world more pleasant and less painful. So, the connection to war is this; war makes the world a worse place. All the violence and killing. All the casualties, both military and civilian. All the shattered lives and families; all the scars left behind. All the property and infrastructure damage. All the costs of fighting the war and having a military, and what could've been done instead with all that money which would have made the world a better place."

David: "Springsteen had an old song, with the line: 'War, what is it good for? Absolutely nothing!'"

Diego: "There you go, he's an American poet as well as a rock star. I was born in Jersey, myself."

David: "So war *always* imposes greater costs than benefits? *Always* creates more pain than pleasure?"

Diego: "I firmly believe that it's true, when you add up everything and take a long-term view."

Elizabeth: "Well, I agree, there certainly have been *many* wars of which that's true: WWI; Vietnam; the Iran-Iraq War. Maybe even *most* wars are like that. But I don't think *all* wars are like that. World War II, especially."

Diego: "Fifty million people died in that war. Fifty million!"

Elizabeth: "I know."

Diego: "And they used atomic bombs, deliberately killing civilians."

Elizabeth: "I know, and I know about the firebombing of Dresden, too."

Diego: "Deliberate civilian destruction, and by the good guys, too. And don't get me started on the bad guys in that war. Just war theory itself, with its rule of discrimination and noncombatant immunity, regards civilians as innocent. War is horribly, and always, wrong because it destroys innocents."

Elizabeth: "But World War II defeated the fascists! In Italy, Germany, and Japan. Completely smashed fascism as a political doctrine and system of government. Think about what an achievement that is; no fascist parties have come to power anywhere in the world since then. *We defeated the kind of people who committed the Holocaust*; doesn't that speak for itself?"

Diego: "At what cost, though?"

Elizabeth: "Well, you just listed the costs of war. Let's consider some benefits. War is often good for the economy, it often spurs technological innovation, and it often unites a nation and convinces it to set aside its internal divisions. War generates social change and shakes countries and cultures out of out-dated practices. Women, for example, first got to vote in wartime circumstances."

Diego: "War is good for women? You've got to be kidding me, Liz . . ."

Elizabeth: "Look, all that I'm saying is that there's costs, and there's benefits. You can't just list all the costs and declare that you've done a complete and comprehensive analysis. The major benefits of a just war, of course, are the moral ones: resisting and defeating aggression; and protecting innocent people who need protection from the aggressors."

Diego: "So, you kill some innocent civilians—theirs—to save others, namely your own? That's a self-serving calculus."

Elizabeth: "It matters who started it. Even kids know this. The aggressor, through his violent rights-violating actions, creates the need for the war. And civilian casualties are unavoidable in wartime. So, to protect your people and stand up for your rights, you've got to do something rough, which you know will hurt people."

Diego: "Not just *hurt* people—*kill* people. Innocent people who *don't* deserve it. And killing innocents is *murder.*"

David: "I don't think it is."

Diego: "What? You can't be serious."

David: "I am. When innocent people—civilians—happen to get killed as a result of your actions, it is not murder if you do not intend for them to be killed. It is only the *deliberate and intentional* killing of innocent people which counts fully as murder."

Diego: "You may not deliberately intend to kill civilians . . ."

Elizabeth interrupted: "Especially since that is, very often, a waste of time in terms of winning a war. It's not rational."

Diego: ". . . as I was saying, you don't *intend* it, but you still *foresee* it. You can predict it will happen. If you are a head of state, you know full well that, if you go to war—even if it's a defensive war against a prior act of aggression—innocent civilians will die. So you *may not do it; you cannot start a process which you know will result in the death of innocent people.*"

David: "Yes you may, provided three things: 1) you may foresee such deaths, but you do not intend them; 2) you respect the laws of war and *jus in bello,* thereby ensuring your side does everything it reasonably can to minimize civilian casualties; and 3) Liz's point remains huge, regarding the fact that it was the aggressor's fault in starting the war in the first place and you've got no choice but to respond with war."

Diego: "Ah, but that's where you're wrong. There are always choices, and there are alternatives to war. Such as nonviolent civil disobedience. You bring up the

Second World War, Liz, but did you know that systematic civil disobedience—
sometimes violent, sometimes peaceful—was used in the Scandinavian countries quite effectively? They held mass strikes, sit-down strikes at the factories,
and when they were forced to work in weapons factories, they deliberately made
mistakes, resulting in shoddy weapons for the Nazis. The Danes and Norwe-
gians, especially. A few resistance fighters even blew up some railway lines and
assassinated some Nazi generals and officials."

Elizabeth: "But, wait, it's not 'nonviolent' resistance if they were shooting officials and using explosives."

Diego: "True, but at least it's not the same *scale* of violence and mass destruction as war. And it made a significant dent in the Nazi war machine."

David: "Well, did it really, Diego? Did it? With all due respect to those people and
those countries, they didn't defeat the Nazis. Systematic non-warring resistance
did *not* defeat the Nazis. What did was war. War, Diego. Not pacifism, not civil
disobedience, whether violent or nonviolent. The Nazis had to be beaten through
war. One of history's greatest and clearest examples of a just war."

Diego: "Let me give you two other examples, of truly nonviolent civil disobedience, and the possibilities it brings regarding the creation of a better world.
Before I do, might I recommend to you both an excellent book, *A Force More
Powerful*, by Jack DuVall and Peter Ackerman? Filled with case studies and devoted to showing that nonviolent resistance to injustice is a force more powerful than violence and war."

Elizabeth: "They made it into a PBS series. I saw it, and thought it was very
good."

Diego: "The first case is Mahatma Ghandi's campaign of nonviolent civil disobedience to force the British out of India during the latter's campaign for national independence in the late 1940s. They used economic boycotts, mass
protests, and systematic noncooperation to make it impossible for the British to
continue colonial rule effectively and profitably. The second case, of course, is
that of the Reverend Martin Luther King, Jr.—inspired by Jesus' example, I
might add—pushing for equal rights for African-Americans in the 1960s. They
used mass protest and demonstrations; they broke the existing laws and were ar-
rested, and fought the charges on principle. They got the media on their side, and
the people reacted when they saw racist Southern cops siccing attack dogs, and
firing water cannons, on nonviolent protesters. The result: The Civil Rights Act
of 1964, and the expansion of equal rights. The use of nonviolent, sustained political protest to bring about an end to injustice. The principled way. The nonviolent, non-waring way."

David: "Don't say: 'Jesus' way.'"

Diego: "I won't, because you just said it for me. Ha! The point remains; you
don't need to resort to violence and war to overthrow unjust social structures
and to effectively resist aggression."

David: "I disagree. Well, wait. It *might* be true in some circumstances, and such strategies certainly do expand our options when we consider how to confront bad guys. *But we can't count on such tactics always being true and effective.* For instance, why did the nonviolent campaigns you mentioned work? With respect, I suggest that it's not just because they were nonviolent and morally superior. There were other factors at play. In the India case, the British were broke at the end of WWII and had to give up their colonies, anyway. In the civil rights case, the media exposure in the northern United States was decisive in the federal government's pushing the southern state governments to accept greater equality. And that's vital; in both these cases, you have an 'Aggressor' who turns out to be morally sensitive, and wants to do the right thing. In the end, the British people didn't, and don't, want to be imperialists and the American people didn't, and don't, want to be racist. So the campaigns of nonviolent resistance reminded them of their true moral commitments, and so change happened. But what happens when the Aggressor is *not* morally sensitive and *won't* be moved or softened by the determined idealism of nonviolent protesters? I mean, c'mon, the Hitlers and Stalins, the Saddams and Maos, of this world are simply not going to respond to such tactics; they are just going to take the opportunity to slaughter the protesters, and take more effective control of their conquests. Again, it was war which beat the Nazis, not nonviolent Norwegian resistance."

Elizabeth: "I'll go further. If a government refuses to go to war to defend its people from violent aggression, it violates one of its most basic duties, namely, to protect and represent its people. For a government to just roll over and die in the face of violent invasion or aggression is an abdication of responsibility, and I think that's why Michael Walzer describes pacifism as 'a disguised form of surrender.' He thinks we should always, out of self-respect, resist rights-violating aggression and he also thinks that, in the final analysis, war remains the last effective resort for restoring and reclaiming your rights against violent aggressors."

Diego: "That seems a bit harsh."

David: "Well, maybe. Perhaps let's put it this way. John Rawls says that pacifism is 'an unworldly doctrine' to believe in. I like that way of putting it more, because it's not as though nonviolent protesters are cowards . . ."

Diego: "Not at all! Would you go to jail for your beliefs? Or let a racist cop beat you, and sic his dog after you, while you deliberately decide to let it happen? That's courage, my friend; bravery in the face of danger."

David: "I agree, so that's why I'm reluctant to agree with Walzer, that it's a form of surrender. But what I think pacifism is, is excessively optimistic about the ability of that form of resistance to always work. In cases where it can, obviously it should be tried prior to war. But there will be cases where it can't work—where the Aggressor is brutal and ruthless—and, in such cases, to still cling to pacifism is naïve and unworldly. It can be a rough-and-tumble world, and we've got to be prepared to defend our lives and our rights with force if need be."

Diego: "That was your turn for eloquence. Good. But then let me push you on some of my prior arguments. Like the whole virtue/vice thing. Obviously, peace is a virtue and war a vice, no? I just mentioned how, in my view, engaging in nonviolent resistance is much more demanding and brave than fighting in war. The way soldiers get lionized in wartime just sickens me."

Elizabeth: "Well, I admit that all the medals, ceremonies, and days of honor can be over-the-top."

Diego: "Yes, and it's the state's way of encouraging other, gullible young men and women to become part of the war machine. Fresh meat for the ravenous Mars, god of war."

Elizabeth: "But war does seem to involve *some* virtue. I would submit that courage and bravery *do* get generated in war. Some tales of battlefield valor are incredible. Soldiers stick together, and fight as a band of brothers, taught to work for something bigger than themselves. That's toughness and commitment, which are virtues. It also takes cleverness to pull off a good battle, or piece of war strategy. But, above all, *a just war resists aggression and protects people who need protecting.* Those things are all part of a good and virtuous life: courage; cleverness; brotherhood; commitment; strength; protecting the weak; resisting and punishing injustice. You can't pretend that war is only about vice and rage and destruction. Look, bottom line: war saved us from a world dominated by the Nazis and the Communists. *Most* wars might be stupid and horrible but *some* wars—just wars—are morally necessary and have improved the world."

Diego: "And now it's your turn for eloquence, Liz. Bravo. But, about the unworldliness thing; who's to say what's naïve and what will work, and what won't? Why should I give up on my ideals for having a peaceful world simply because you people assure me I'm being naïve? Should I just surrender my ideals and say: 'OK, let's drop a bomb on them? Kill a whole bunch of them?' I mean, don't you two want to live in a world one day devoid of war?"

David: "Of course."

Elizabeth: "Yes. I just don't see that world coming into being at any time in the near future. We must thus be prepared."

Diego: "Whereas I say that the only way ever for such a world to come into being is precisely by risk-taking idealists clinging to their values and refusing to cave in. That is how the world truly progresses. The majority say: 'you can't do that; that's not possible,' but then the idealistic minority replies 'we'll just see about that.' And they go and transform the world for the better."

David: "I hope they do, Diego. Truly. It's just that history gives us no confidence in the belief that war is going to go away. So responsible political leaders, in the meantime, must plan prudently on having to defend the lives and rights of their people, by force if necessary."

Elizabeth: "Let me ask you this, Diego. Some of my colleagues in the military say that pacifists undermine national security because their criticism of war undermines soldier morale. They actually think pacifists are a kind of threat, I guess. A subversive threat. I've also heard it commonly said that pacifists are selfish freeloaders: they get a free ride on the security provided by others—and then have the audacity to look down their noses at those who provide them with that security."

Diego: "Being in the military, I come across such accusations all the time. They are nonsense. First, we live in a free society based on human rights. These things allow us to speak our minds freely. All too often in the past, people have appealed to national security to silence and suppress opinions they fear and don't like to hear. If a soldier truly believes in a war, my criticism of war won't undermine his morale. But it may get him to question, and think through, his beliefs. It may even make him think twice about whether the latest war truly is good, or whether the politicians are lying. And how can such an increase in critical self-reflection be bad for a free society?"

David: "And the freeloader argument?"

Diego: "Also known as the 'clean hands argument,' the pacifist just selfishly wants to keep his own hands clean for the sake of his inner moral purity. He lets others do his dirty work for him, then looks down on them. It is a very unflattering portrait."

David: "And we know it doesn't describe you, Diego. Still, what would be your response?"

Diego: "My response is that, when you look at pacifism historically, you see that pacifists are not stuffy, selfish, holier-than-thou jerks. On the contrary, they have been thrown in jail, they have been spat upon and called cowards, they have been ostracized and denied career opportunities, and so forth. They are not selfish pigs; they just sincerely believe that peace is the way to a better world, and not just for themselves, but for everyone. I mean, get real; was Ghandi only looking out for himself? Did Martin Luther King, Jr. look down his nose at others? Were all those protesters selfish pigs? No, they were just trying to make the world better, both for themselves and others."

David: "Good."

Diego: "But we're not done."

David: "No?"

Diego: "No. We must return to this issue of killing innocents. We cannot just brush it under the table; it's absolutely central and vital. It's the most important ethical issue."

David: "I agree."

Diego: "You laid out a series of conditions according to which killing the innocent could be justified."

David: "Yes, you're permitted to foresee it, intellectually, but not to intend it or want it, from the point of view of your will. Also, during war, you scrupulously observe the rules of war to respect civilians and to minimize their casualties. Finally, and most importantly, the cause for launching the war is just, and vitally important to secure—like defeating an aggressor like Nazi Germany."

Diego: "Well and good, from a *theoretical* point of view. But let me ask you this, from a *practical* point of view; have these conditions actually ever been satisfied in the real world? Take your favorite example of WWII. A good cause, on the part of the Allies, I admit. Germany was the aggressor in first invading Poland, and the Nazis represented a horribly unjust political system. But then there are real problems, in my view, with the other two; the Allies didn't scrupulously observe the rules of war, and sometimes they intentionally attacked civilians. Like when the United Kingdom deliberately bombed the residential areas of Germany, and the United States dropped nukes on Hiroshima and Nagasaki. My concern, then, is this; you specify nice theoretical points, but in the real world, war becomes so nasty that none of your precious just war rules can ever fully be met, and so, *even by your own principles,* every war is unjust. Thus, why don't you come over to my side and become a pacifist?"

David: "An excellent argument. What I like about just war theory is that, in my view, it is a middle-ground compromise between the extremes of realism and pacifism. It is not too pessimistic, like the realists, nor too optimistic, like the pacifists. And it allows for more complex judgments. Pacifists make sweeping claims, as you just did, about war in general. I don't like that; I like instead how just war theory judges the beginning, middle, and end of each war separately. Thus I can agree with you that big mistakes were made by the Allies in WWII in terms of violating *jus in bello* and civilian targeting. But I can still claim that, in terms of *jus ad bellum*, the war was a just war to fight and had a great cause; the defeat of fascism."

Diego: "But wait, doesn't just war theory say that, unless *all* its rules are satisfied, the war as a whole must be deemed unjust? So, if the rules of *jus in bello* were violated, the Second World War was unjust."

David: "No: *one aspect* of the war was unjust; others were just. There's the separation, at least conceptually, between the cause of the war and the conduct during the war."

Diego: "But that's my point. In the real world, the rules of just conduct during war always get violated, the process grows corrupt and out of control, and innocent people get killed. These tragedies are not—repeat, NOT—redeemed by the fact that they were committed in the name of a good cause. *The bad conduct corrupts the good cause,* and renders the war on the whole unjust. Since civilian casualties happen in every war, even ones with good causes, it follows that war in our world will forever be unjust."

Elizabeth: "But you can't ignore the cause, Diego. The cause is the whole reason for the war and, in my view it sets the tone for everything else; the middle and end. I agree with you that, if there's no good cause for the war, then the whole business from start to finish is corrupt and unjustified. Garbage in, garbage out. But, if there is a good initial cause to the war, that changes and establishes the context for how we should fight and how we should end wars."

Diego: "I know, I know. You just war theorists have what I call a 'top-down' perspective: the cause at the start determines everything else, and you're willing to tolerate some civilian casualties for the sake of securing that cause. Pacifists, by contrast, have a 'bottom-up' perspective: we view war as a certain kind of act or process, namely, one which murders innocents. Civilian killing is intrinsic to the process of war. Since that is the nature of warfare, we are not willing to tolerate any civilian casualties for the sake of the supposedly grand or great cause. And this difference of opinion leads to two very different judgments about war."

David: "How can we determine which view prevails?"

Diego: "Well, I'm just shocked at how you cannot agree with my perspective; when you act, your action itself must be free from moral violation. If not, it doesn't matter the cause or end you're trying to achieve. Both the cause and the conduct have to be violation-free for the whole business to count as justified."

Elizabeth: "Whereas I can't see how you'd be willing to lay over and die, not resisting aggression, simply because you know that, if you do, there will be some civilian casualties. It's wrong to let aggressors have their way with the world."

Diego: "First, I didn't say not to resist; I just said we cannot use *war* as a tool of resistance. We should use instead systematic nonviolent civil disobedience."

David: "But we've just been down that road. It won't always work. And then where are you?"

Diego: "It may work more than you think. And, if it doesn't, then at least you know you didn't commit murder in order to save yourself."

David: "That is supposed to be a comforting thought to take to your grave? Plus, again, Diego, please watch the rhetoric; it's not murder if it's not intentional and if it serves a good cause and if you take many substantial precautions to minimize the chances of civilian casualties happening."

Diego: "Look, I agree with Socrates' principle: *it is better to suffer injustice than to inflict it oneself.*"

Elizabeth: "But, as David said, is that supposed to be a comforting thought while the aggressor takes over your country, and either kills or enslaves you?"

Diego: "Now who's using rhetoric! And yes, why shouldn't that be comforting, to know you went to your grave nonviolently resisting aggression?"

Elizabeth: "Because maybe if you *did* use violent resistance, you wouldn't be going to the grave at all! You'd be beating back the aggressor, and preserving the freedom of your society."

Diego: "Potentially, but again at the cost of having gotten your hands dirty in this way. Me, I'd rather keep my hands clean and, if it means the aggressor wins and kills me, I'll meet my Maker with a good conscience."

David: "But does it ultimately boil down to that, Diego? And I mean this very seriously; can pacifism only be plausible in a religious context? You're willing to stick with nonviolence because you believe that, even if you do get killed, there's an afterlife and God will reward you for your nonviolent restraint."

Elizabeth: "Whereas if you're like me, and you believe that this life is all we have, and there is neither God nor an afterlife, you're going to want to hang on to that one life and defend it with force if need be."

Diego: "Maybe it does, I don't know. I've always linked my pacifism with my religion so it's hard for me to think of them separately."

David: "But then my point would be that you've failed to give me, in the end, a secular reason to believe in pacifism."

Diego: "No, I've given you several, and good ones at that. You just don't want to believe them because, in the end, you just war theorists are, like the realists, too cynical and accepting of war. You're too complacent about the status quo. Contrast that attitude with the inspiring optimism and determination of the pacifist protesters."

David: "I'd like to think that's not true. My own work on postwar social reconstruction is motivated by a desire to change the world for the better and make it less violent and more respecting of human rights."

Nick suddenly rounded the corner, and approached the table where Elizabeth, David, and Diego were debating. His eyes were red and puffy, his facial expression etched with deep sadness.

Nick: "My nephew Gil has died on the operating table. They couldn't save him."

Elizabeth: "I'm so horribly sorry, Nick."

She stood and hugged him. Then David and Diego did the same.

Nick: "Can you believe it? Dead. My poor little nephew. I can still see him as a kid playing in our backyard when they would visit. Now dead, here in a foreign country, and because of a foreign war. I just want to go home. Liz, I'm sorry, but

Lori and I must collect Gil's body and return it back home to America. The Pentagon will contact you today about getting a new escort, or whether you even need to go forward to Baghdad at this point."

Elizabeth daubing tears: "Thank you, Nick. Again, I'm so sorry. Please tell Lori the same thing from me."

At that moment, Lori rounded the corner. She looked utterly devastated: empty, disheveled, exhausted.

Lori: "I heard you myself, Liz. Thank you."

Hugs were had, all around. During his hug, Diego said to Lori: "Gil has gone to a better place, into the arms of his loving heavenly Father."

Lori breathed silently: "Thank you. I can only hope that's true. What Gil does have, for sure, is peace and rest from the suffering of his injuries. The peace of the grave, brought about by fighting in an unjust war."

Nick: "No, a just war. Just a much more difficult one than what we expected. Success might still be possible. Gil was part of a great team, trying to do great things, and make the world truly safer and better. We can only hope that effort succeeds."

Lori: "Well, I hope it does. Truly. I don't want Gil's death to have been pointless or, even worse, in the service of a war that only made the world worse off. But it's hard to be optimistic now, right? My love lies dead on a slab. My hope for a future family and happiness. Gil has peace, but I don't. I wonder when, or whether, I ever will."

THE END

Bibliography

Anscombe, G. E. M. "War and Murder," in R. Wasserstrom, ed., *War and Morality* (Belmont, CA: Wadsworth, 1970), 41–53.

Aristotle. *The Politics*, trans. by C. Lord (Chicago: University of Chicago Press, 1984).

Augustine. *The City of God*, trans. by R. Ryson (Cambridge: Cambridge University Press, 1998).

Bentham, Jeremy. *An Introduction to the Principles of Morals and Legislation* (London: Dover, 2007).

Clausewitz, Carl von. *On War*, trans. by B. Heuser (Oxford: Oxford University Press, 2007).

Dobbins, J. et al. *America's Role in Nation-Building: From Germany to Iraq* (Washington: RAND, 2003).

Dobbins, J. and S. Jones, eds. *The United Nations' Role in Nation-Building* (Washington: RAND, 2006).

Doyle, Michael. *Ways of War and Peace* (New York: Norton, 1997).

Du Vall, Jack and Peter Ackerman. *A Force More Powerful* (New York: St. Martin's, 2000).

Falk, Richard. *The Costs of War* (London: Routledge, 2007).

Faludi, Susan. *The Terror Dream* (New York: Metropolitan Books, 2007).

Ferguson, Niall. *Colossus* (New York: Basic Books, 2004).

Freud, Sigmund. *Civilization and Its Discontents* (New York: Norton, 2005).

Gelven, Michael. *War and Existence* (Philadelphia: Penn State University Press, 1994).

Grotius, Hugo. *The Laws of War and Peace*, trans. by L. R. Loomis (Roslyn, NY: WJ Black, 1949).

Hobbes, Thomas. *Leviathan* (London: Touchstone, 1997).

Huntingdon, Samuel. *The Clash of Civilizations* (New York: Simon and Schuster, 1990).

Kant, Immanuel. *Perpetual Peace and Other Essays*, trans. by T. Humphrey (Indianapolis: Hackett, 1983).

Keegan, John. *A History of Warfare* (New York: Vintage, 1994).

Mannix, Daniel. *A History of Torture* (London: The History Press, 2003).

Mill, John Stuart. *The Basic Writings of John Stuart Mill* (New York: Modern Library, 2002).

Niebuhr, Reinhold. *Christianity and Power Politics* (New York: Charles Scribner's Sons, 1940).

Nietzsche, Friedrich. *The Will to Power*, trans. by W. Kaufman (New York: Vintage, 1968).

Porter, Bruce. *War and the Rise of the State* (New York: The Free Press, 2002).

Rawls, John. *A Theory of Justice* (Cambridge, MA: Harvard University Press, 1971).

Sen, Amartya. *Development as Freedom* (London: Anchor, 2000).

Sun Tzu. *The Art of War*, trans. by S. Griffith (New York: Oxford University Press, 1963).

Vitoria, Fransisco de. *Political Writings*, eds. A. Pagden and J. Lawrence (Cambridge: Cambridge University Press, 1991).

Walzer, Michael. *Just and Unjust Wars* (New York: Basic Books, 1977).

Weber, Max. *General Economic History* (London: Dover, 2003).

Welch, D. *Justice and The Genesis of War* (Cambridge: Cambridge University Press, 1995).

Woodward, Bob. *Plan of Attack* (New York: Simon and Schuster, 2004).

About the Author

Brian Orend is professor of philosophy and director of international studies and global engagement at the University of Waterloo in Canada. He is the author of five books, including the widely used *Human Rights: Concept and Context* and *The Morality of War*. Orend lectures around the world on the ethics of war and peace, and is also the president of LIB Publishing, Inc. His Ph.D. is from Columbia University in New York City.

Breinigsville, PA USA
08 June 2010
239446BV00002B/8/P